T0023527

Quantum physicist, *New York Times* bestselling author, and BBC host Jim Al-Khalili reveals how 8 lessons from the heart of science can help you get the most out of life

Today's world is unpredictable and full of contradictions, and navigating its complexities while trying to make the best decisions is far from easy. *The Joy of Science* presents 8 short lessons on how to unlock the clarity, empowerment, and joy of thinking and living a little more scientifically.

In this brief guide to leading a more rational life, acclaimed physicist Jim Al-Khalili invites readers to engage with the world as scientists have been trained to do. The scientific method has served humankind well in its quest to see things as they really are, and underpinning the scientific method are core principles that can help us all navigate modern life more confidently. Discussing the nature of truth and uncertainty, the role of doubt, the pros and cons of simplification, the value of guarding against bias, the importance of evidence-based thinking, and more, Al-Khalili shows how the powerful ideas at the heart of the scientific method are deeply relevant to the complicated times we live in and the difficult choices we make.

Read this book and discover the joy of science. It will empower you to think more objectively, see through the fog of your own preexisting beliefs, and lead a more fulfilling life.

"Jim Al-Khalili has distilled the very essence of science. This book is packed full of joy, inspiration, and real wisdom."

Alice Roberts, Professor of Public Engagement in Science, University of Birmingham

"Jim Al-Khalili eloquently reminds us of all the reasons to celebrate science. A lovely little book that will serve you well as a trusted guide in this troubled post-truth era."

Sabine Hossenfelder, physicist and author of *Lost in Math*

"*The Joy of Science* pulls back the curtain on the essential nature of science and tackles the confusions the public confronts in understanding how it's done. I highly recommend Al-Khalili's book to anyone, scientist or not, interested in thinking more scientifically."

S. James Gates Jr., coauthor of *Proving Einstein Right*

"In the age of post-truth politics, when misinformation and conspiracy theories flood social media and endanger lives, Al-Khalili's book is a patient, gentle, and humane corrective. *The Joy of Science* is a call for a more rational and discerning attitude to what we experience in our lives, guided by respect for expertise and critical judgement but also by compassion."

Philip Ball, author of *Beyond Weird* and *Curiosity*

"Jim Al-Khalili is justly celebrated as a leading expositor of science. In this book, he distills the nature and limits of our scientific knowledge and highlights how the scientific mindset can help us in everyday life. His wise precepts are especially welcome at a time when, despite science's triumphs, public discourse is bedeviled more than ever by fake news and conspiracy theories. We'd all be better citizens if we took his message to heart—this book deserves wide readership."

Martin Rees, author of *On the Future*

"Science is a way of thinking about and understanding the world—and in this captivating book, Al-Khalili argues that we all should be thinking more scientifically. Writing exquisitely about the complexities of scientific concepts and ideas, he uncovers our biases and dispels common myths and misunderstandings about how the world and science works. His highly entertaining book is crucial reading for all of us, especially at this time of global pandemic and climate crisis, when finding solutions depends critically on a deeper understanding of what science is and isn't."

Sarah-Jayne Blakemore, author of
Inventing Ourselves

"This is a beautiful, straightforward, and readable little book with so much to say about how and why we do science. I recommend it to anyone in these crazy times who wants to understand the meaning and value of following the science."

Daniel M. Altmann,
Imperial College London

"This pithy and insightful book provides readers with a collection of fun and timely ideas in an accessible way."

Sean Carroll, author of
Something Deeply Hidden

"Jim Al-Khalili's latest masterpiece beautifully conveys just how profound, intimate, and unique our connection with science is. *The Joy of Science* awakens the scientific thinking that is deeply rooted in all of us, revealing not only what its methods truly are but also how one can find enlightenment by trying them out."

Claudia de Rham,
Imperial College London

"Al-Khalili's timely and inspirational writing allows us all to experience a touch of the 'joy' of science."

Helen Pearson,
Chief Magazine Editor of *Nature*

THE JOY OF SCIENCE

THE JOY OF OF SCIENCE

JIM AL-KHALILI

PRINCETON UNIVERSITY PRESS
PRINCETON AND OXFORD

Published by Princeton University Press
41 William Street, Princeton, New Jersey 08540
99 Banbury Road, Oxford OX2 6JX

press.princeton.edu

All Rights Reserved

Library of Congress Cataloging-in-Publication Data

Names: Al-Khalili, Jim, 1962- author.
Title: The joy of science / Jim Al-Khalili.
Description: Princeton : Princeton University Press, [2022] |
    Includes bibliographical references and index.
Identifiers: LCCN 2021029263 (print) | LCCN 2021029264 (ebook) |
    ISBN 9780691211572 (hardback) | ISBN 9780691235660 (ebook)
Subjects: LCSH: Science—Philosophy.
Classification: LCC Q175 .A485 2022 (print) | LCC Q175 (ebook) |
    DDC 501—dc23
LC record available at https://lccn.loc.gov/2021029263
LC ebook record available at https://lccn.loc.gov/2021029264

British Library Cataloging-in-Publication Data is available

Editorial: Ingrid Gnerlich and Whitney Rauenhorst
Production Editorial: Mark Bellis
Text and Cover Design: Chris Ferrante
Production: Jacquie Poirier
Publicity: Sara Henning-Stout and Kate Farquhar-Thomson
Copyeditor: Annie Gottlieb

This book has been composed in Adobe Text and Futura PT

Printed on acid-free paper. ∞

Printed in the United States of America

10 9 8 7 6 5 4 3 2 1

For my father

CONTENTS

PREFACE

As a young student in the mid-1980s, I read a book called *To Acknowledge the Wonder* by the English physicist Euan Squires. It was about the latest ideas in fundamental physics (at the time), and I still have it somewhere on my shelf nearly four decades later. While some of the material in that book is now outdated, I have always liked its title. At a time when I was contemplating a career in physics, the chance to 'acknowledge the wonders' of the physical world was what really inspired me to devote my life to science.

There are many reasons why people pursue their interests in one subject or another. In science, some enjoy the thrill of climbing into the crater of a volcano or crouching on a cliff's edge to observe birds nesting—or looking through telescopes or microscopes to see worlds beyond our senses. Some design ingenious experiments on their laboratory workbenches to reveal the secrets inside stars, or build giant underground particle accelerators to probe the building blocks of matter. Some study the genetics of microbes so they can develop drugs and vaccines to protect us against them. Others become fluent in mathematics and scrawl pages upon pages of abstract but beautiful algebraic equations, or write thousands of lines of code that instruct their supercomputers to simulate Earth's weather or the evolution of galaxies, or even model the biological processes inside our bodies. Science is a vast enterprise, and there is inspiration, passion, and wonder everywhere you look.

But the old adage that beauty is in the eye of the beholder applies to science as well as more generally in our lives. What we regard as fascinating or beautiful is highly subjective. Scientists know as well as anyone that new subjects and new ways of thinking can be daunting. When you haven't been properly introduced to a subject, it can seem downright forbidding. However, my response would be that, if we try, we can almost always gain a better understanding of an idea or concept that might once have seemed unfathomable to us. We just have to keep our eyes and minds open and take the time we need to think

things through and absorb the information—
not necessarily to the level of experts, but just
enough to comprehend what we need.

Let's take as an example a simple and common phenomenon in the natural world: the
rainbow.[1] We can all agree that there is something enchanting about rainbows. Is their magic
diminished if I explain to you the science of how
they form? The poet Keats claimed that Newton
had "destroyed all the poetry of the rainbow, by
reducing it to the prismatic colours." In my view,
far from 'destroying its poetry,' science only en-
hances our appreciation of nature's beauty. See
what you think.

Rainbows combine two ingredients: sun-
shine and rain. But the science behind the way
in which they combine to create the arc of colour

1 In beginning this book by invoking the iconic rainbow,
I am traveling a path well-worn by other science writers; for
example, Carl Sagan (*The Demon-Haunted World: Science as
a Candle in the Dark*) and Richard Dawkins (*Unweaving the
Rainbow: Science, Delusion and the Appetite for Wonder*). I
hope readers already familiar with these books will make al-
lowances for my having followed in this tradition for the benefit of new readers who may be coming afresh to the example.

we see in the misty sky is as beautiful as the sight itself. Rainbows are made of broken sunlight that reaches our eyes after the Sun's rays strike a billion raindrops. As the Sun's rays enter each water droplet, all the different colours of light that make up sunlight slow down slightly to travel at different speeds, bending and separating out from each other in a process called refraction.[2] They then bounce off the backs of the droplets, returning to pass through their fronts at different points, refracting a second time as they do so and fanning out into the colours of the rainbow. If we measure the angles between the sunbeam and the different-coloured rays that emerge from the veil of raindrops in front of us—we find that they range from 40 degrees for violet light, which undergoes the most refraction and so forms the innermost colour of the rainbow, to 42 degrees for

---

2 Sunlight, or white light, is made of different colours, each of which has a different wavelength. When it encounters a medium, like air or water, it slows down; but each of its constituent colours slows down by a different amount, depending on its wavelength, which causes each colour to have a different angle of refraction.

red light, which undergoes the least and forms the outer rim of the rainbow (see the diagram).[3]

Even more wondrously, this arc of splintered sunlight is really just the top part of a circle—the curved surface of an imaginary cone lying on its side, whose tip is located in our eyes. And because we are standing on the ground, we only see the top half of the cone. But, if we were able to float up into the sky, we would see the entire rainbow as a complete circle.

You cannot touch a rainbow. It has no substance; it does not exist in any particular part of the sky. A rainbow is an intangible interaction between the natural world and our eyes and brains. In fact, no two people see the same rainbow. The one we see is made from those rays of light that have entered our eyes alone.

3 The type of rainbow I have described is called a primary rainbow. We can sometimes also observe outer, fainter secondary rainbows, which are produced when the Sun's rays undergo two, rather than one, internal reflections within each raindrop. In these cases, we see only the rays of colour that emerge at angles of between 50 and 53 degrees. But in secondary rainbows, due to this double reflection, the colours are reversed, with red on the inside and violet on the outside.

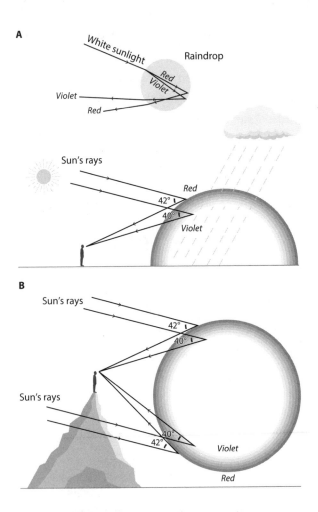

A

White sunlight

Raindrop

Red

Violet

Violet

Red

Sun's rays

Red

42°

40°

Violet

B

Sun's rays

42°

40°

Sun's rays

40°

42°

Violet

Red

DIAGRAM OF A RAINBOW

So, each of us experiences his or her own unique rainbow, created by nature for us and us alone. This, for me, is what a scientific understanding can give us: a richer, more profound—and more personal—appreciation of the world; one we would never had had without science.

Rainbows are so much more than just a pretty arc of colour, just as science is so much more than hard facts and lessons in critical thinking. Science helps us see the world more deeply, enriches us, enlightens us. My hope is that this book will welcome you to a world of light and colour, truth and profound beauty—a world that will never fade as long as we all keep our eyes and minds open and share what we know with each other. The closer we look, the more we can see and the more we can wonder. I hope you will join me in acknowledging the wonder—the joy of science.

THE JOY OF SCIENCE

INTRODUCTION

As I write these words in the spring of 2021, and while we all continue to reel from the impact of the Covid pandemic, we are witnessing a seismic shift in the way people around the world view science: its role and value to society, how scientific research is carried out and its claims tested, and indeed how scientists conduct themselves and communicate their discoveries and results. In short, and albeit in the most devastating and tragic of circumstances, science and scientists are today under scrutiny like never before. Certainly, the race to understand the SARS-CoV-2 virus and to find ways of defeating it have highlighted the fact that humanity cannot survive without science.

Though there will always be those who fear science and treat it with suspicion, I see among the vast majority of the world's population a new appreciation for and trust in the scientific method, as ever more people realize that the fate of humanity rests not so much in the hands of politicians, economists or religious leaders, but in the knowledge that we gain about the world through science. Equally, scientists are coming to appreciate that it is not enough to keep the findings of our research to ourselves. We must also put in the effort to explain, as honestly and transparently as possible, how we work, what questions we ask and what we have learned, and to show the world how our newly discovered knowledge can best be put to good use. Today, in a very real sense, all our lives depend on the thousands of virologists, geneticists, immunologists, epidemiologists, mathematical modellers, behavioural psychologists and public health scientists around the world working together to defeat a deadly microscopic organism. But the success of the scientific enterprise also depends on the public's willingness, both collectively and as individuals, to make informed decisions for ourselves, as well as for our loved ones and the broader societies we live in, that make good use of that knowledge gained by scientists.

The continuing success of science—be it in tackling the biggest challenges facing humanity in the twenty-first century, such as pandemics, climate change, eradicating disease and poverty, or in creating wondrous technologies, sending missions to Mars and developing artificial intelligence, or simply learning more about ourselves and our place in the universe—all depends on a relationship of openness and collaboration between scientists and non-scientists. This can only happen if politicians pull back from the all-too-prevalent current attitudes of isolationism and nationalism. Covid-19 is no respecter of national borders, cultures, race or religion. None of the biggest problems facing us as a species is. Therefore, just like scientific research itself, tackling such problems must also be a collective, collaborative enterprise.

Meanwhile, nearly eight billion human inhabitants on the planet still have to navigate through their daily lives, make decisions and act on them, often while stumbling through a dense fog of confusing information . . . and misinformation. How then can we take a step back and see the world, and ourselves, more objectively? How can we sort through all the complexity and do better for ourselves and for each other?

The truth is that complexity isn't new. Misinformation and confusion aren't new. Huge gaps in our knowledge aren't new. The world we face is daunting, confusing, even overwhelming at times. None of this should be news to us, of course. In fact, science is built on this very premise; humans came up with the scientific method precisely to deal the difficulties of making sense of a confusing and complex cosmos. In our daily lives, every one of us—scientists and non-scientists alike—encounters a world bursting with information, which is constantly reminding us of our own ignorance. What can we do about it? Indeed, *why* should we do something about it?

In this book, I have put together a short, all-purpose guide to thinking and living a little more scientifically. Before reading on, you might take a moment to ask yourself this: Do I want to find out about the world as it really is? Do I want to make decisions based on that knowledge? Do I want to mitigate a fear of the unknown with a sense of promise, potential, and even excitement? If you are tempted to say "yes" to any of the above questions, and even if (or dare I say, especially if) you don't yet know how you feel about them, then maybe this book can help.

As a practicing scientist I do not profess to impart any profound wisdom, and I certainly hope there is no hint of superiority or condescension in the tone of this book. My aim is simply to explain how thinking scientifically can offer you some control over the complex and conflicting information that the world throws at you. This book does not contain lessons in moral philosophy, nor a list of life skills or therapeutic techniques to help you feel happier or more in control of your life. What I have to say comes from the core of what science is and the ways in which it is practiced: an approach that is tried and tested and that has served humankind well over our centuries-long quest to understand the world. Yet, at a deeper level, the reason it has served us so well is that it was built to help people like you and me make sense of complexity or gaps in our knowledge, and generally to arm us with a confidence and a better sense of perspective when we encounter the unknown.

Because the way we do science has served humankind so well, for so long and so successfully, I think it is worth sharing with you this way of thinking.

Before I present my case for why we should all be thinking more scientifically, I need to say something about how scientists themselves think. Scientists are as embedded in the real world as everyone else, and there are ways of thinking shared by all scientists that everyone can follow when encountering the unknown and making decisions in daily life. This book is about sharing these ways of thinking with everyone. They have always been for everyone, but somewhere along the line, that fact seems to have been lost.

Firstly, contrary to what many people think, science is *not* a collection of facts about the world. That is called 'knowledge'. Rather, science is a way of thinking and making sense of the world, which can *then* lead to new knowledge. There are, of course, many routes to gaining knowledge and insight, whether through art, poetry and literature, religious texts, philosophical debate, or through contemplation and reflection. That said, however, if you want to know about how the world really is—what physicists like me sometimes refer to as the 'true nature of reality'—then science has a big advantage, for it relies on the 'scientific method'.

THE SCIENTIFIC METHOD

When we talk about the scientific 'method', it implies there is just one way of 'doing' science. This is wrong. Cosmologists develop exotic theories that explain astronomical observations; medics carry out randomised control trials to test the efficacy of a new drug or vaccine; chemists mix compounds together in test tubes to see how they react; climatologists create sophisticated computer models that mimic the interactions and behaviour of the atmosphere, oceans, land, biosphere, and Sun; while Einstein figured out that time and space can bend in a gravitational field by solving algebraic equations and doing a lot of deep thinking. While this list hardly scratches the surface, there is a common theme running through it. One could say that all of the above activities involve a curiosity about some aspect of the world—the nature of space and time, the properties of matter, the workings of the human body—and a desire to learn more, to reach a deeper understanding.

But isn't this too general? Surely, historians are curious too. They too look for evidence in order to test a hypothesis or uncover some previously unknown fact about the past. Should we then regard history as a branch of science? And what about the conspiracy theorist who claims that the Earth is flat? Is he or she not just as curious as a scientist, just as keen to find rational evidence that supports a claim? Why then would we say that they are not being 'scientific'? The answer is that, unlike scientists, or indeed historians, flat-Earth conspiracy theorists would not be prepared to reject their theory when presented with irrefutable evidence to the contrary, such as NASA images from space showing our planet's curvature. Clearly, just being curious about the world does not mean someone is thinking scientifically.

There are a number of features that distinguish the scientific method from other ideologies, such as falsifiability, repeatability, the importance of uncertainty and the value of admitting mistakes, and we will be considering each of these throughout this book. But, for now, let us look briefly at a few features that the scientific method shares with other ways of thinking—ways that we would not necessarily regard as proper science—in order to show that no single one of these features *alone* is sufficient to meet the rigorous requirements of the scientific method.

In science, one should continue to test and question a claim or hypothesis even when there exists overwhelming evidence supporting it. This is because scientific theories need to be *falsifiable*—that is, a scientific theory must be capable of being proved false.[4] To offer a classic

4 In the philosophy of science, a theory is falsifiable (or refutable) if it can be contradicted or disproved by evidence, whether that be in the form of observations, laboratory measurements or mathematics and logical reasoning. The idea was introduced by the philosopher Karl Popper in the 1930s.

example, I could put forward a scientific theory that all swans are white. This theory is falsifiable, since you could prove its falsehood by observing just one swan of a different colour. If evidence is found to contradict my theory, then the theory must be either modified or discarded. The reason conspiracy theories are not proper science is because no amount of contrary evidence would dissuade their advocates. In fact, a true conspiracy theorist sees any evidence as bolstering his or her preexisting views. In contrast, a scientist takes the opposite approach. We change our minds in the light of new data, because we are trained to shun the absolute certainty of the zealot who insists that only white swans exist.

A scientific theory also needs to be testable and held up to the light of empirical evidence and data. That is, we should be able to use a scientific theory to make predictions, and then see if those predictions are borne out in experiments or observations. But again, this is not enough on its own. After all, an astrological chart also makes predictions. Does that make astrology a real science? And what if the prediction made by the astrological forecast comes true? Does that give it the seal of approval?

Let me tell you the story of the faster-than-light neutrinos. Einstein's special theory of relativity, which he published in 1905, predicts that nothing in the universe can travel faster than light. Physicists are now so confident this is true that they generally insist there must be a mistake if a measurement shows that something *is* moving faster than light. But this is exactly what was reported in 2011 in a now-famous experiment involving a beam of subatomic particles called neutrinos. Most physicists did not believe the results. Was this because they were being dogmatic and closed-minded? A layperson may well think so. Contrast this with the astrologer who claims that your stars will be aligned on Tuesday and you will receive good news, which sure enough comes true when your boss offers you a promotion. In one case, you have a theory conflicting with experimental data, and in the other, you have a theory whose prediction is borne out by events. How then can we say that relativity is a valid scientific theory and yet astrology is not?

As it turned out, physicists were right not to give up so easily on relativity theory, because the team that had carried out the neutrino experiment soon discovered that a fibre optic cable had been attached improperly to their timing device, and fixing it eliminated the faster-than-light results. The fact is that if this experiment *had* been correct and neutrinos do indeed travel faster than light, then thousands of other experiments that proved the contrary would have had to be wrong. But there was a rational explanation for the surprising experimental results, and the theory of relativity held firm. Yet we trust it not because it survived refutation by an (ultimately wrong) experimental result, but because so many other experimental results have confirmed the correctness of the theory. In other words, the theory is falsifiable, and it is testable, and yet it continues to stand strong, fitting in with so much of what we know to be true about the universe.

In contrast, a correct astrological prediction is sheer luck, since no physical mechanism could possibly explain it. For instance, since the astrological signs were invented, the view of the sky has changed due to a shift in Earth's axis; so, you were not born under the sign you likely thought you were anyway. More importantly, our modern astronomical understanding of the true nature of stars and planets has rendered any theoretical basis for assigning meaning to astrological signs useless. In any case, if astrology *were* true, and distant stars, whose light takes many years to reach us and whose gravitational effects are far too weak to be felt on Earth, could influence future events within the mind-bogglingly complex affairs of humankind, then this would mean that all of physics and astronomy would have to be discarded, and we would need a new, irrational and supernatural explanation for all the phenomena that science currently explains so well and on which the modern world, including all of its technology, is built.

Another feature of the scientific method that one often hears is that science is *self-correcting*. But since science is just a process—a way of approaching and seeing the world—it is wrong to think that this implies science itself has some kind of agency. What the statement really means is that *scientists correct each other*. Science is carried out by people. And we all know that people are fallible, especially since, as we've discussed, the world is a complex and confusing place. So, we test each other's ideas and theories, we argue and we discuss, we interpret each other's data, we listen, we modify, we extend— sometimes we give up entirely on an idea or experimental result if other scientists, or even we ourselves, show it to be flawed. Crucially, we see this as a strength, not a weakness, for we don't mind being proven wrong. Naturally, we want our own theories or interpretations of the data to be correct, but we don't cling to them when there is strong evidence to the contrary. If we're wrong, we're wrong, and we cannot hide from that—and it would be embarrassing to even try. That's why we do our best to subject our own ideas to the toughest criticism and tests we can think of before we announce them, and even then, we 'show all our workings' and we quantify our uncertainty. After all, even if we've looked everywhere for a black swan and haven't seen one, that does not mean there isn't one out there somewhere that we simply have not found yet.

When it comes to deciding whether or not something is 'proper' science, I am not claiming that there is a list of criteria against which to judge it—boxes to tick off in order to differentiate between science and non-science—for there are plenty of examples scattered throughout science that do not adhere to one or more of the criteria of the scientific method. I can immediately think of several examples in my own field of physics. Is superstring theory—the mathematical idea that all matter is composed of tiny strings vibrating in higher dimensions—not proper science because we don't (yet) know how to test it and therefore cannot claim it to be falsifiable? Is the Big Bang theory and the expansion of the universe not proper science because it is not repeatable? The enterprise of science and how we do it is far too broad to be neatly packaged, and it should not be considered as something hermetically sealed away, separated from other pursuits, such as history, art, politics or religion. This book is not about articulating separations or detailing distinctions, nor is it about uncovering the faults and shortcomings of the scientific method. Rather I aim to distil what is best about science and its method, and how it can be used as a power for good if applied to other walks of life.

There are, of course, many ways in which scientific research carried out in the real world can be improved. For example, if mainstream science is predominantly carried out, and its validity is decided upon, by white men in the Western world, doesn't this mean it is tainted, even shaped by certain prejudices, whether intentionally or unintentionally? Surely, if there is little or no diversity of views, and all scientists see, think, and question the world in a similar way, then they will not, as a community, be as objective as they maintain they are, or at least aspire to be. The solution is that there should be far greater diversity in the practice of science: in gender, ethnicity and social and cultural backgrounds. Science works because it is carried out by people who pursue their curiosity about the natural world and test their and each other's ideas from as many varied perspectives and angles as possible. When science is done by a diverse group of people, and if consensus builds up about a particular area of scientific knowledge, then we can have more confidence in its objectivity and truth. A democratized science can help to protect against the emergence of dogma, whereby an entire community of scientists in a particular field accepts a set of assumptions or ideas as being absolute without ever questioning them further, to the extent that dissenting voices are suppressed or dismissed. However, there is an important distinction to be made between dogma and consensus, for sometimes the two can be confused. Established scientific ideas have earned the right to be widely accepted and trusted, even though they could one day be improved upon or replaced, because they have so far survived the myriad and diverse questions and tests to which they've been subjected.

'FOLLOWING THE SCIENCE'

Sociologists will argue that to truly understand how science works we need to embed it in the broader contexts of human activities, whether they be cultural, historical, economic or political.

To simply talk about 'how we do science' from the perspective of a practitioner like me is, they would say, too naïve, for science is more complicated than that. They will also insist that science is not a value-neutral activity since all scientists have motives, biases, ideological stances and vested interests, just like everyone else, whether it is to secure a promotion, enhance a reputation or establish a theory they have spent years developing. And even if the researchers themselves don't have biases or motives, then their paymasters and funders will. Needless to say, I find such an appraisal overly cynical. While those who carry out the science, or indeed those who pay their salaries, will almost inevitably *not* be value-free, the scientific knowledge that they gain *should be*. And this is because of the way the scientific method works: self-correcting, building on firm foundations of what has already been established as factually correct, being subject to scrutiny and falsification, reliant on reproducibility, and so on.

But then I would say that, wouldn't I? After all, I want to persuade you of my own objectivity and neutrality. And yet I too cannot be entirely objective, nor value-free, however much I may think I am or try to be. But the subjects I study— the theory of relativity, quantum mechanics or the nuclear reactions taking place inside stars—are all value-neutral descriptions of the external world, as are genetics, astronomy, immunology and plate tectonics. The scientific knowledge we have gained about the natural world—the description of nature itself—would be no differ- ent if those who had discovered it spoke differ- ent languages, or had different politics, religions or cultures—provided, of course, that they are honest and truthful and carry out their science well and with integrity. Of course, our research priorities—the questions we might ask—depend on what is considered important at that time in history or in that part of the world, or on who- ever has the power to decide what is important and what (and whose) research to fund; these decisions can be culturally, politically, philo- sophically, or economically driven. For exam- ple, physics departments in poorer countries are more likely to fund research in theoretical physics than experimental physics since laptops and whiteboards are cheaper than lasers and particle accelerators. These decisions about what questions to pursue and what research to fund can also be subject to bias; and so the more diversity we can foster among those in positions of leadership and power the more the scientific enterprise can protect itself against bias when determining which veins of research are more or less promising or potentially impactful. All this said, what is ultimately learned about the world—the knowledge itself, achieved by doing good science—should not depend on who has carried out that science. A scientist located at an elite institution may reach a different result from a scientist located at another institution that is not regarded as elite; but one has no inherent claim on a more accurate result than the other. By the nature of science and the accumulation of evidence, the truth will out.

Many who are suspicious of the motives of scientists argue that science, as a process, can never be 'value-free'. To some extent, as we've discussed, they are correct. However much we scientists might think that our pursuit of knowledge and truth is objective and pure, we must acknowledge that the ideal of *all* science being value-free is a myth. Firstly, there are values *external* to science, such as ethical and moral principles about what we should or should not be studying, and social values, such as public interest concerns. Such external values must play a role in the decisions about what science should be funded and conducted—and, of course, those decisions can be subject to bias, which we must be mindful of and work against. Secondly, there are values *internal* to science, such as honesty, integrity and objectivity, which are the responsibility of the scientists carrying out the research. This is not to say that scientists should not also have a say in shaping or debating those external values, for they have a responsibility to consider the consequences of their research, both in terms of how it may be applied and in terms of the policies it might shape and the public's reaction to it. Sadly, all too often scientists will argue among themselves as to whether science can in principle be value-free, confusing the value-free pursuit of pure knowledge about the world—in astrophysics, for example—with the inevitably value-laden research in fields such as environmental science or public health policy.[5]

But assuming that we can agree that science in the real world is not entirely value-free, and that the knowledge gained through the process of good science is, let us go on to explore a few of the challenges that the public sometimes has with the perception of science, both justified and unjustified.

Scientific progress has undoubtedly made our lives immeasurably easier and more comfortable. With the knowledge that has been revealed through science, we have been able to cure diseases, create smartphones and send space missions to the outer Solar System. But this success can sometimes have the adverse effect of giving people false hope and unrealistic expectations. Many can be so blinded by the success of science

5 For an excellent account of this issue, see the book by Heather Douglas, *Science, Policy, and the Value-Free Ideal* (Pittsburgh: University of Pittsburgh Press, 2009).

that they will believe any report or marketing trick that sounds remotely 'scientific', whatever the source and however bogus the product may be. This is not their fault, for it is not always straightforward to tell the difference between real scientific evidence and misleading marketing based on unscientific notions.

Understandably, most people tend not to worry too much about the scientific process itself, only about what science can achieve. For example, when scientists claim to have discovered a new vaccine, the public wants to know if it is safe and it works, and they will either trust that the scientists involved know what they are doing, or they will be suspicious (of the scientists' or their paymasters' motives). Chances are, it will only be other scientists in the field who will delve into whether the research was carried out at a reputable lab, whether the vaccine has been through rigorous randomised clinical control trials, and whether the research is published in a reputable journal and has been through the proper peer review process. They will also want to know if the results claimed are repeatable.

It also doesn't help the public to make up their minds about what or whom to trust when scientists disagree, or when they express uncertainty about their results. While this is perfectly normal in science, many people nevertheless wonder how they can believe anything scientists say if the scientists themselves are never quite sure. Not properly communicating the importance of uncertainty and debate in science is one of the main problems we face today when explaining how we develop our scientific understanding of the world.

It can become even more confusing for the public when the advice—particularly on issues relating to public health—is not only conflicting, but reaching them from sources outside the scientific community, such as the media, politicians, online posts, or after having been spread over social media. In reality, even genuine scientific discoveries reach the public after having been through a number of filters, whether the lab or university press officer who has had to distil a simplified message from a complex scientific paper, the journalist looking for a headline, or the amateur science enthusiast who posts information online. This might range from what precautions to take during a pandemic, to the risks of vaping or the benefits of flossing. And as the story develops and spreads, so too will opinions about it—both informed and uninformed—so that we end up mostly believing what we want to believe anyway. Instead of making careful, evidence-based rational judgements, many people will accept something as true if it fits in with their preconceived prejudices and ignore what they don't want to hear.

Before I move on, I should also say a few words about the advice scientists give to governments, the purpose of which is to inform policy decisions. While scientists can provide all the evidence they have, from the results of laboratory experiments or computer simulations, clinical trial data, graphs and tables, to the conclusions they are able to draw from their results, in the end what is done with this scientific advice is down to the politicians. I should make clear that scientists should always advise on the basis of their specific area of expertise. Thus, epidemiologists, behavioural scientists and economists may all have views on what is best for the population when fighting Covid, and the politicians must then weigh up costs and benefits of what may sometimes be conflicting advice. An epidemiologist might estimate the number of excess deaths due to Covid associated with delaying going into lockdown by one week, while an economist might calculate that that delay avoids loss to GDP which might lead to an equivalent or greater number of deaths. Both experts will have based their conclusions on model predictions that may well be highly accurate given the data and model parameters used, and yet they predict different conclusions. It is then the role of policymakers and politicians to choose what they regard as the best course of action. The public also has choices to make. The more individuals in a population are given access to those conclusions in a transparent way, and take up the challenge of learning to understand them, the more they will be empowered to make informed choices—in daily life and as part of the democratic process—that will benefit them and their loved ones.

Science, unlike politics, is not an ideology or belief system. It is a process. And we know that politicians base policy decisions on more than just the scientific evidence. So even if the science is clear-cut, when it comes to the complexity of human behaviour, decision-making is never value-free. Nor, I have to admit with some reluctance, should it be.

Politicians, like most people, almost always follow the science that aligns with their preferences and ideologies. They will cherry-pick the conclusions that fit their purposes, often influenced by public opinion, which is in turn shaped by how the facts are presented in the media or official government guidelines or by scientists themselves in the first place. Basically, the relationship between science, society and politics involves complex feedback loops. And lest you think I am being overly critical of politicians, I am the first to acknowledge that scientists are not elected, and it is therefore not our job as scientists to say what policies should be put in place. All we can do is communicate as clearly as possible and provide guidance based on the best scientific evidence available at the time. We may personally feel very strongly about an issue, but that should not colour the advice we give. In a democracy, whether we support a particular government or not, it is the elected politicians in the end who have to make the decisions and be held accountable for those decisions, not the scientists—although there is no doubt that society would benefit immeasurably if we had more scientifically trained politicians, and more scientific literacy generally.

Luckily, this book is not about the complicated relationship between science, politics and public opinion, but about how we can import the best features of the scientific process into our wider decision-making and opinion-forming processes in daily life. The scientific method is a combination of curiosity about the world, a willingness to question, to observe, to experiment and to reason, and of course to modify our views and learn from experience if what we discover does not follow our preconceived thinking.

Here then is a brief guide to how we can all think and behave more rationally. Each chapter is a piece of advice, distilled from some particular aspect of the scientific method. We may find that sharing a more scientific approach to thinking about the world can lead us to a better place.

1

SOMETHING IS EITHER TRUE, OR IT ISN'T

How many times have you gotten into an argument with a friend, colleague or family member—or, even worse, with a stranger on social media—and stated what you thought was a clear-cut fact, only to hear the response, "Well, that's your view", or "That's one way of looking at it"? These responses—often polite, sometimes aggressive—are examples of the insidious and disturbingly common phenomenon of 'post-truth'. Defined by the Oxford Dictionary as *"relating to or denoting circumstances in which objective facts are less influential in shaping public opinion than appeals to emotion and personal belief,"* post-truth has become so prevalent that the term was 'word of the year' in 2016. Have we moved too far away from objective truth, to the extent that even proven facts about the world can be conveniently dismissed if we don't like them?

Even while we find ourselves in a postmodern world of cultural relativism, the internet, and social media in particular, is driving society towards ever-increasing polarisation of opinions on all manner of cultural and political issues, and we are expected to pick sides, with each one making a claim on the 'truth'. When a blatantly untrue assertion, motivated by a particular ideological belief, holds sway over an undeniable fact or over knowledge supported by reliable evidence, we see the phenomenon of post-truth politics in action. On social media, it is most often seen linked to conspiracy theories or in the pronouncements of populist leaders or demagogues. Sadly, this irrational way of thinking has infected many people's attitudes more generally, including their views towards science, and we often see claims on social media that opinion is more valid than evidence.

In science, we use different models to describe nature; we have different ways of building up our scientific knowledge, and we regularly create different narratives depending on what aspects of a phenomenon or process we want to understand, but this is not the same as saying that there are alternative truths about the world. A physicist like me tries to uncover ultimate truths about how the world *is*. Such truths exist independently of human feelings and biases. Gaining scientific knowledge is not easy, but acknowledging that there *is* a truth out there, towards which we can strive, makes our mission clearer. Following the scientific method, critiquing and testing our theories and repeating our observations and experiments, ensures that we can get closer to that truth. But even in our messy everyday world, we can still adopt a scientific attitude to reach the truth of a matter—to help us see through the fog. We must therefore learn to spot and weed out what are 'culturally relative' truths or ideologically motivated truths and examine them rationally. And when we encounter the sorts of falsehoods referred to as 'alternative facts', we must remember that those who advocate them are not trying to present a believable narrative to replace an original fact, but merely to create plausible levels of doubt to suit their ideologies.

There are many situations in everyday life in which acknowledging the existence of objective truth and taking steps to seek it out can prove to be of far greater value than convenience, pragmatism or self-interest. How then do we get to this truth—not my truth or your truth, not conservative truth or liberal truth, not Western truth or Eastern truth, but *the* truth about something—however trivial? And to whom can we turn for help? How can we be sure of a source's honesty and objectivity?

Sometimes it is easy to see why a person, group or organisation holds a particular view, for they may have a particular motive or vested interest. For example, if a representative of the tobacco industry tells you that smoking is not really harmful and that the health risks are exaggerated, then you should rightly dismiss what they say. After all, they would say that, wouldn't they? But all too often people mistakenly apply this same reasoning when they don't need to. For example, if a climatologist says that Earth's climate is changing rapidly and that we need to modify our lifestyles to prevent catastrophic consequences, a climate change denier will often counter with, "Well, of course they would say that. . . . They're in the pay of 'x'" (where 'x' could be an environmental group or green energy company, or just perceived liberal academia).

I am not denying that in certain cases this cynicism may be justified, for we can all think of examples of research that is funded for ideologically driven or profit-driven motives. And we must also be wary of so-called data dredging— also known as 'p-hacking'—whereby analysis of data is misused deliberately in order to find something that can be presented as statistically significant, then only reporting those cherry-picked conclusions.[6] I will say more about this in chapter 6 when I discuss confirmation bias.

6 See for example, M. L. Head et al., "The extent and consequences of p-hacking in science", *PLoS Biology* 13, no. 3 (2015): e1002106, doi:10.1371/journal.pbio.1002106.

But these inevitable biases notwithstanding, a suspicion of science or denial of its findings often occurs because of a misunderstanding of how science works.

In science, an explanation that has survived the scrutiny of the scientific method can become an established fact about the world, adding to our cumulative scientific knowledge . . . and that fact is not going to change. Let me give you my favourite example from physics. Galileo came up with a formula that allowed him to calculate how quickly an object falls when dropped. But his formula was more than 'just a theory'. We still use it over four centuries later because we know it to be true. If I drop a ball from a height of five metres, it will fall for one second[7] before it hits the ground—not two seconds or half a second, but

7 In fact, it will take a little over one second (more like 1.01 seconds), and the precise value will depend on where on Earth's surface I drop the ball, since the gravitational pull on a falling object can vary ever so slightly from one place to another depending on local geology, altitude above sea level and even how far we are from the equator, since the Earth is not a perfect sphere.

one second. This is an established, absolute truth about the world that is never going to change.

In contrast, when it comes to the complexity of individual human behaviour (psychology) or the way humans interact within society (sociology) we find, inevitably, that there is more nuance and ambiguity. This suggests that there can indeed often be more than one 'truth', depending on how we see the world. This is not the case when it comes to the physical world, such as the time it takes for a ball to fall to the ground. When natural scientists, such as physicists, chemists or biologists, state that something is either true or it isn't, they are not talking about complex moral truths, but about objective truths about the world.

To show you what I mean, I will present a list of randomly selected facts, each of which is either true or false. They are not debatable or subject to opinion, ideological belief or cultural background, and we can use the scientific method to confirm or dismiss each of them. The conclusions we draw about them will also not change over time. Some readers may wish to dispute a few—maybe by saying something like, "But that's just your opinion", or "How can you possibly be so certain? I thought the scientific method always left room for doubt", and so on. However, the items on this list are meant to show that while we must always be open to new ideas and explanations in science and that what we once thought to be true may turn out not to be so once we gain a deeper understanding, we do know some things for sure. We really do. The reason I am being so bullishly confident is because if science is wrong about any one of the items in the list below, then the whole edifice of scientific knowledge would need to be pulled down and rebuilt. Even worse, all the technology that relies on that knowledge would have been impossible to create. And I find that so exceedingly unlikely that I am as near to certain about them as one can be in science. Anyway, here's the list:

1. Humans have walked on the Moon—*True*
2. The Earth is flat—*Not True*
3. Life on Earth evolved through a process of natural selection—*True*

4. The world was created about six thousand years ago—*Not True*
5. Earth's climate is changing rapidly, mainly due to humanity's actions—*True*
6. Nothing can travel through space faster than the speed of light in a vacuum—*True*
7. Give or take, there are roughly seven billion billion billion atoms in the human body—*True*
8. 5G masts contribute to the spread of viruses—*Not True*

For each of these examples, I can provide mountains of evidence to support their truth or falsehood. But that would be very boring. On the other hand, what is more interesting to explore is why some people might disagree with me if, I would argue, they are *not* thinking scientifically. Take the idea of falsifiability. The philosopher Karl Popper stated that we can never *prove* a scientific theory to be correct since that would require us to test it in every conceivable way, which is impossible. However, a single counter-example can prove a theory false. You'll remember this from the example of the white swans I mentioned earlier. Popper argued that the idea of falsifiability was a crucial feature of the scientific method. However, a weakness of his argument is that the counterexample offered—for example, an experimental result—might itself be false. Maybe the brown swan that refutes the claim that all swans are white is simply caked in mud. This was the case in the famous faster-than-light neutrinos experiment I mentioned in the introduction. Unfortunately, it is precisely this loophole that conspiracy theorists fall back on in denying the validity of any evidence against their pet theory—whether it is the claim that the Moon landings were a hoax, or that the Earth is flat or that the MMR vaccine causes autism in children. They will forever claim that the evidence against their theory is itself false. This is a classic example of misusing one of the tools of the scientific method—by denying and rejecting any evidence that falsifies one's theory, never offering a rational scientific reason for that rejection, nor ever stating what form of evidence one would demand as sufficient to falsify one's theory.

The opposite scenario is even more fascinating: when something factually true is denied *in spite of* overwhelming evidence. This denial can take several forms: the most basic is called *literal denial*, and it means just that: a simple refusal to accept or believe the facts. Then there is *interpretive denial* in which the facts are accepted but interpreted differently to fit in with the person's ideology, culture, politics or religion. Finally, and most interesting of all, is *implicatory denial*, coined by the sociologist Stanley Cohen.[8] This states that if A implies B and I don't like B, then I will reject A too. For example, the theory of evolution implies that life evolves randomly and without purpose. But this goes against my religious beliefs, so I reject evolution theory. Or: acting to stop climate change requires me to change my lifestyle, which I am not prepared to do, therefore I reject the claims that the climate is changing or that we can do anything about it.

8 This idea is outlined in Cohen's book *States of Denial: Knowing About Atrocities and Suffering* (Cambridge, UK: Polity Press, 2000), where he discusses all the ways in which uncomfortable realities are avoided and evaded.

Or: to stop the spread of the Covid virus we must follow Government advice, stay at home and lose income, and wear face masks when out in public. These restrict my fundamental freedoms and therefore I reject the scientific evidence that calls for such actions.

Of course, there is a world of difference between hard scientific facts and the sorts of messy, vague truths we encounter in everyday life. When a particular statement about something is embedded within the complex morass of beliefs, feelings, behaviours, social interactions, decision-making or any of the other million issues we encounter and debate, then it can often be more complicated than simple black and white. This does not mean that the statement is untrue, but rather that, by itself, it may not be entirely valid in all situations. Even a simple statement can be both true and false depending on the context; it can be true in one situation, but not another. In some cases, the same can be true in science too. When I stated the fact that a ball dropped from a height of five metres will hit the ground after one second, I failed to mention the context in which it is a true fact: namely that this only applies on Earth. A ball dropped from five metres above the surface of the Moon will take almost two and a half seconds to hit the ground because the Moon is smaller than the Earth and so has a weaker gravitational pull. It's the same scientific formula that we use—that is an absolute truth—but the numbers we plug in to get the answer are different. Sometimes, even scientific truths have to be put into context.[9]

A simple truth can also be expanded to include more information and give us a deeper understanding, which can take it in a different direction. For example, the scientific fact about how long a ball takes to hit the ground, whether on Earth or on the Moon, is explained by New-

9 If you wish to find out more about the nature of truth, then you should read the work of the late philosopher of science Peter Lipton. For example, his 2004 Royal Society Medawar Lecture, "The truth about science", *Philosophical Transactions of the Royal Society B* **360**, no. 1458 (2005): 1259–69, https://royalsocietypublishing.org/doi/abs/10.1098/rstb.2005.1660), or his article "Does the truth matter in science?" in *Arts and Humanities in Higher Education* **4**, no. 2 (2005): 173–83, doi:10.1177/1474022205051965.

ton's law of gravity. But we now have a deeper and more profound picture of the nature of gravity thanks to Einstein's theory of relativity. While the time taken for a ball to fall is a fact (given the context) that will never change, we now have a better understanding of what is going on. Newton's picture of gravity as an invisible force pulling the ball to the ground has been replaced with Einstein's picture of masses bending spacetime around them (I won't go into the physics here, but if you are interested, I have written several nontechnical accounts).[10] And even this deeply profound picture may one day be replaced with a more fundamental theory of gravity; but the fact about how long it takes for the ball to hit the ground is not going to change.

It is all very well, you might be thinking, to come up with examples from science in which a truth can depend on the context, but how does this manifest itself in our everyday world? Well, here is an example: the statement "More exercise

10 For example, my recent book, *The World According to Physics* (Princeton University Press, 2020).

is good for your health" is, you might argue, indisputable, but it is not true if you already exercise too much or have a medical condition that makes certain exercise dangerous.

There are those who argue that personal and cultural biases, societal norms and historical contexts should be taken into account when deciding whether or not something is true. The theory known as social constructivism holds that truth is constructed by social processes; in fact, that all knowledge is 'constructed'. This means that our perception of what is true is also subjective. It's an idea that has even influenced our scientific representations of reality, such as the definition of race, sexuality and gender. Sometimes, there is a valid and important point to be made. However, taking these arguments too far can ultimately lead us towards the dangerous idea that truth is whatever we as a society decide to agree upon. That, I'm afraid, is nonsense.

This is certainly not how most scientists view the world. On the whole, science has progressed, and our knowledge of the physical universe expanded, thanks to what is known as scientific realism, which states that science provides us with an increasingly accurate map of reality that is independent of our subjective experience. In other words, there are facts about our universe that are true regardless of how we decide to interpret them, and if we have more than one interpretation of what is going on, then that is our problem to resolve, not the universe's. It may be that we are never able to find the right interpretation of what is going on and the best we can hope for is an explanation that satisfies all the criteria of a good scientific theory: for example, that it explains all existing evidence as well as making new testable predictions we can measure and check it against. Or maybe we have to wait for future generations to come up with a better theory or interpretation, just as Einstein's explanation of gravity replaced Newton's. My point is that scientists know that even if our current *understanding* of some aspect of physical reality is hazy, that does not mean that the very existence of a real world is up for debate.

So, how does the idea that there are objective scientific truths about the world help us in deciding or arguing about whether capitalism is good or bad, or whether abortion is right or wrong? Let us briefly consider what at first sight would seem to be clear-cut moral 'truths' and see if we can use rational arguments to test their objectivity. Here are four statements:

1. Showing kindness and compassion is a good thing
2. Murder is wrong
3. Human suffering is bad
4. Actions more likely to bring harm than benefits are bad.

On first reading, you might think that none of these statements is contentious. Surely, they are all examples of universal and absolute moral truths. However, each of them must be seen in context. Consider the first statement. It could be argued that this is just a tautology, and you might as well say that being good is good. So, in a sense, it is meaningless. How about the second statement that murder is wrong? What if you'd had a chance to kill Hitler before the Holocaust? Is murdering one man right if you *knew* it could prevent millions of innocent deaths? For the third, regarding human suffering, what about guilt or grief? They too are forms of suffering. Are they also bad? Should we try to avoid *all* suffering if we can, or should we embrace some kinds because they give our lives meaning? And for the last one, often an action or decision can bring benefits to some, yet harm to others. So, who decides which outweighs the other?

You can see that, while there are many moral truths we might initially regard as obvious, it is not difficult to pick holes in them if we really want to (as indeed we see on social media when someone says something that they regard as perfectly reasonable). Also, moral truths that are desirable for us to accept and abide by are different from scientific truths, such as a ball taking one second to fall to earth. Despite this, most of us can agree that there *do* exist universal moral traits and standards of human behaviour that span time and cultures and that all humanity should at least try to follow and practice, such as compassion, kindness and empathy. These qualities may well have evolved in humans and higher

mammals because they conferred an evolutionary advantage, but that does not make them any
less desirable now that society has developed to
the extent that they are no longer necessary for
our survival. As for the above four statements,
dismantling them does not require inventing
counterfactual scenarios. Just embedding them
in contexts where they do not apply is enough
to show that they are not absolute. Yet this
does not make them untrue in other contexts.
It just means that moral truths need to be well
framed—just like the scientific fact that it takes
a ball one second to fall a distance of five metres
needs to be well framed by specifying that it is
only true on Earth.

Many of the issues we grapple with in everyday life are messy. Often, two diametrically
opposed views about some issue can each be
based on a fundamental truth, because each is
valid within its own domain of applicability. I
guarantee that many opinions you hold may not
be straightforwardly true or untrue, but rather
are based on a core of truth along with myriad
assumptions, misconceptions, biases, guesses,

wishful thinking and/or exaggerations. However, if you are prepared to put in the effort, you can often sieve all of these away and leave behind just the plain facts of a matter—the nuggets of truth and the bare falsehoods. You can then see how to formulate a better-informed opinion in answer to a question. Thinking like a scientist means learning to study issues objectively—to break each down to its constituent parts, to look at it from different angles, but also to zoom out to gain a wider perspective.

Many of us do this already, of course, across various walks of life—from police detectives trying to solve a case, to investigative journalists uncovering a political scandal, or doctors diagnosing an illness. In all these professions the scientific method is being applied to analyse a problem and discover the hidden truth. While such people are all highly trained and skilled in what they do, we can all, maybe to a lesser extent, apply the same basic philosophy to our lives. So, don't just accept what you see or what you're told. Analyse it carefully, break it down, take into account all reliable evidence and consider all possible options.

Despite all of humankind's faults and frailties, biases and confusion, there are still facts out there about the world—objective truths that exist whether or not someone believes them. Don't let anyone tell you otherwise.

2

IT'S MORE COMPLICATED THAN THAT

We are told that the simplest explanations are usually the right ones. After all, why complicate something more than you need to? This assumption is often applied in everyday life, where unfortunately it is not always true. The idea that simple explanations are more likely to be correct than complicated ones is known as Ockham's razor—named after the English medieval monk and philosopher William of Ockham.

A well-known example of the use of this principle in science was the overthrow of the geocentric model of the universe developed by the ancient Greeks, in which the Sun, Moon, planets and stars all orbit the Earth, which itself sits in the centre of the cosmos. The core principle of the model was the aesthetically appealing notion that all the heavenly bodies moved around us in perfect concentric spheres. This picture held sway for two thousand years, despite the fact that it gradually became more cumbersome and complicated as it tried to account for the observed motion of planets like Mars, which was seen to slow down, speed up, and sometimes even double back on itself.[11] To rectify this 'retrograde' motion, additional smaller circular paths followed by some of the planets, called epicycles, were bolted onto the primary orbits to ensure that the model still accurately matched astronomical observations. Later, other additions to the model, such as shifting Earth slightly away from the centre about which all the other bodies orbited, were also included. Then, in the sixteenth century, Nicolaus Copernicus swept away this makeshift

---

11 We now know that this is just a result of our view of Mars from Earth. Both Mars and Earth orbit around the Sun at different distances and different speeds. Earth is a little faster because it is closer to the Sun. This means that a Mars year is 687 Earth-days long.

model and replaced it with his much simpler and more elegant heliocentric picture in which the Sun and not the Earth is at the centre of the universe. Both the Earth-centred and Sun-centred models 'worked' in the sense that they predicted the motions of heavenly bodies, but we now know that only one of them is correct, and it is the cleaner, simpler Copernican one— the one without all the clumsy extras. This, we are told, is Ockham's razor in action.

But the above account is wrong. While Copernicus correctly replaced the Earth with the Sun at the centre of the known cosmos, he still believed the planetary orbits to be perfectly circular, rather than the *less* 'elegant' elliptical orbits we now know them to be thanks to the work of Kepler and Newton. And so, he did not in fact get rid of the unwieldy patch-ups and extensions to the old geocentric model, as he still needed them for his heliocentric model to work. Although we now know that the Earth does indeed go round the Sun and not the other way round, we also know from modern astronomy that the true dynamics of the Solar System are far more complex than anything the ancient Greeks could have imagined—Ockham's razor in reverse.

An equally famous example in the history of science is Darwin's theory of evolution through natural selection, which explains the mind-boggling variety of life we find on Earth, all of which evolved over a span of billions of years from one single origin. Darwin's theory is based on a few simple assumptions: 1) that individuals within a population of any species vary; 2) that these variations pass down the generations; 3) that more individuals are born in each generation than can survive; 4) that those with better adapted characteristics to suit their environment are more likely to survive and reproduce. That's it. The theory of evolution is simple. However, wrapped up in these modest assumptions are the mind-bogglingly complex fields of evolutionary biology and genetics, which are among the most challenging areas in all of science. In any case, if we are to truly apply Ockham's razor to the complexity of life on Earth, then surely the non-scientific theory of creationism—that all life was

created as it is today by a supernatural creator—
is far simpler than Darwinian evolution.

The lesson here is that the simplest explanation
is not necessarily the correct one, and the correct
one is often not as simple as it first appears. Ockham's razor, as applied in science, does not mean
that a new theory should replace a previous one
just because it is simpler or because it has fewer
assumptions. I prefer a different interpretation of
Ockham's razor: that a better theory is one that
is more *useful*, because it makes more-accurate
predictions about the world. Simplicity is not
always what we should strive for.

In everyday life too, things are often not as
simple as we would like them to be. To para-
phrase Einstein, we should try to make things
as simple as possible, but no simpler. Nevertheless, the idea that simpler is better seems to have
spread, and we can see a trend towards simplis-
tic arguments, particularly in relation to ethical
or political issues, which ignore all subtlety and
complexity, reducing everything to the lowest
common denominator, distilling the issues into
memes and tweets in which all nuance is lost.

It is certainly tempting, when trying to make sense of a messy world, to reduce a complex issue down to a clean and unambiguous viewpoint—forgetting that there is more than one way to simplify complexity, depending on what aspects you choose to downplay or to emphasise. This means that often, two or more completely diverging views can condense out of a single complex issue, with each then regarded as an incontestable truth by its proponents. But, like much of science, real life is messy, and all sorts of factors and considerations need to be taken into account before we make up our minds about something. Unfortunately, too many people these days are not prepared to put in the effort to look a little deeper, beyond the superficial. Keep it simple, they say, don't blind me with details. And yet, it can be surprising how much clearer, and simpler to understand, an issue becomes if we actually acknowledge its complexity and examine it from different perspectives.

This idea is one with which physicists are familiar. We say that something is *reference frame dependent*. Thus, a ball thrown out of a moving car's window will appear to be travelling at different speeds depending on the frame of reference of the observer—for example, an occupant of the car or someone observing from the side of the road. There is no *absolute* value for the speed of the ball, so both the car occupant and the external observer are right when they quote their different measured speeds of the ball. They are correct in their own frames of reference. Sometimes, what one can say about something depends on perspective and scale. The world as seen and experienced by an ant is very different from that of a human, or an eagle or a blue whale. Likewise, the observations of an astronaut in space differ from those of her fellow humans on the ground.

This dependence of how we see the world on our frame of reference can make it harder to find out how the world really 'is'. In fact, many scientists and philosophers would argue, correctly, that it is impossible to know reality as it actually is, since we can only ever say how we *perceive* it: the way our minds interpret the signals from our senses. But the external world exists independently of us, and we should always try our best to find ways of understanding it that are not subjective—that are reference frame *independent*.

Simplifying an explanation, description or argument is not always a bad thing to do. In fact, it can be very useful. To truly understand a physical phenomenon, to reveal its essence, the scientist will attempt to strip away the unnecessary detail and expose its bare bones (always 'as simple as possible, but no simpler'). For example, laboratory experiments are often carried out under specially controlled conditions to create artificial and idealised environments that make the important features of a phenomenon easier to study. Unfortunately, this hardly ever applies when it comes to human behaviour. The real world is messy and often far too complicated to simplify. There is a well-known joke—to physicists, at any rate—about a dairy farmer who wishes to find a way of increasing the milk production of his cows and so seeks the help of a team of theoretical physicists. After carefully studying the problem, the physicists finally tell him they have found a solution, but that it only works if they assume a spherical cow in a vacuum.[12] Not everything can be made simpler.

A few years ago, I interviewed Peter Higgs, the British physicist after whom the famous particle[13] is named, for my BBC radio programme, *The Life Scientific.* I asked him if he could explain what the Higgs boson was in thirty seconds. He looked at me solemnly and, I have to admit, not particularly apologetically, and shook his head. He explained that it had taken him many decades to understand the physics underlying the Higgs mechanism in quantum field theory, so what right did people have to expect such a complex subject to be condensed into a short soundbite? There is a

---

12 Spherical objects are much easier to describe mathematically than more complicated, cow-shaped ones, and doing laboratory experiments inside a chamber from which all the air has been extracted (a vacuum) means there is less of a chance that the air can influence the results, especially if the experiment involves particles so tiny that they can be buffeted by collisions with the air molecules.

13 The Higgs boson is a short-lived elementary particle that was predicted to exist by a number of theoretical physicists in the 1960s, including Peter Higgs. It was finally detected in particle collisions at the Large Hadron Collider in CERN, Geneva, in 2012.

similar story about the great Richard Feynman who, on receiving his Nobel Prize in the mid-sixties, was asked by a journalist if he could explain what his prize-winning work was in a single sentence. Feynman's response is legendary: "Hell! If I could explain in a few words what it was all about, it wouldn't be worth no Nobel Prize!"

It is human nature to look for the simplest account of something we don't understand, and if we do find a simple explanation, we hang on to it because of its strong psychological appeal over more complicated explanations that we may not be willing to invest the effort to fully understand. Scientists are no different—even the best of us. Soon after Einstein completed his General Theory of Relativity in 1915, he applied its equations to the description of the evolution of the entire universe. However, he found that his equations predicted a universe that was collapsing in on itself due to the mutual gravitational pull of all the matter it contains. Einstein knew that the universe didn't appear to be collapsing, and the simplest assumption he could make was that it had to be stable. So, he modified his equations and chose the easiest mathematical 'fix' possible: he added a number, known as 'the cosmological constant', which did the job of counteracting the part of his equations describing the cumulative attractive gravitational pull of matter, and so he stabilised his model of the universe. But it didn't take long for other scientists to suggest a different explanation: What if the universe wasn't stable after all? What if it was actually getting bigger and all gravity was doing was slowing down its expansion, rather than causing it to collapse? This explanation was confirmed by the astronomer Edwin Hubble in the late 1920s. Einstein realised then that there was no longer any need for his 'fix'. He got rid of his cosmological constant, calling it the biggest blunder of his life.

Jump forward to the present day, however, and we find that scientists have now reinstated Einstein's fix. In 1998, astronomers discovered that the universe is not only expanding, but that its expansion is accelerating. Something is counteracting the cumulative gravitational pull of matter, causing the universe to expand ever faster. We call that something, for want of a better name, 'dark energy'. This is a good example of how our scientific understanding can grow as new evidence and new knowledge accumulates. The fact is, based on what was known a century ago, Einstein chose the simplest solution. But he chose it for the wrong reason. He assumed the universe was static—not expanding or collapsing. Today, it seems the cosmological constant may be necessary to describe our universe after all, but for reasons more complex than Einstein could have realised. And this is not the end of the story, for we still don't really understand dark energy.

Scientists therefore try not to be seduced by Ockham's razor. The simplest explanation is not necessarily the right one. It is a lesson we would do well to carry over into everyday life. We are currently living in an age of soundbites, slogans and instant access to news and information, which has coincided with a move towards more-strident and uncompromising opinions. Society is becoming increasingly ideologically polarised, with complex issues that require open debate and thoughtful analysis being reduced to black or white. All shade is lost, leaving just two opposing views, with the antagonists unwavering in their certainty that they are right. In fact, anyone daring to highlight that an issue is more complicated than either side wishes to admit can find themselves attacked by both sides—if you're not 100 percent with me then you are against me.

What if we applied a little of the scrutiny and cross-examination that is the hallmark of the scientific method to political and social issues that we feel strongly about? Einstein admitted his blunder when he discovered the universe wasn't behaving as simply as he thought. Like science, everyday life is not always simple, as stressed by the title of the best-selling book by science writer Ben Goldacre.[14] Just because we *want* to have simple solutions to problems doesn't mean they are the best ones, or even that they exist, and simple arguments aren't always the right way to understand complex issues.

One often hears it said that such-and-such must be true because it is obvious, or that it stands

14 Ben Goldacre, *I Think You'll Find It's a Bit More Complicated Than That* (London: 4th Estate, 2015).

to reason that it is so, or that it is simply common sense. Scientists learn that explanations of natural phenomena which we might regard as straightforward—obvious even—are not necessarily the correct ones. What we call common sense, to quote Einstein yet again, is nothing more than the accumulated prejudices we have acquired early in life. Thinking that something is true because we have a simple explanation for it is an unreliable way to proceed. Before making up our minds about an issue, we would do well to learn a lesson from Einstein. To avoid serious blunders, jettison your assumptions and invest a little more effort in exploring further. OK, Einstein could not have predicted the existence of dark energy—that had to wait for powerful telescopes that could pick up images from the edge of the universe. But often the truth of a matter is there for us to find with far less effort than that required to discover dark energy. If you are prepared to dig a little deeper, you will be rewarded. Not only will your view of the world become richer, but your outlook on life will be more fulfilling.

3

MYSTERIES ARE TO BE EMBRACED, BUT ALSO TO BE SOLVED

One of my favourite TV shows as a teenager was a series called *Arthur C. Clarke's Mysterious World*. This was a thirteen-part British television series looking at all manner of unexplained events, weird phenomena and urban myths from around the world, introduced by the famous science fiction writer and futurist Arthur C. Clarke. The series divided up its subject matter into three categories of mysteries.

Mysteries of the First Kind are phenomena which were inexplicable and baffling to our ancestors but are now well understood, mostly thanks to the knowledge we've gained through modern science. Obvious examples include natural phenomena like earthquakes, lightning and pandemics.

Mysteries of the Second Kind involve phenomena that are yet to be explained, but which we are confident have rational explanations that we hope to find one day. These phenomena are only mysteries because we have yet to understand them. Examples might include the original purpose of Stonehenge, the prehistoric circle of giant stones in Wiltshire in England; or in physics, the nature of dark matter, the invisible substance that holds galaxies together.

Mysteries of the Third Kind include phenomena for which we have no rational explanation, nor can we see how we could ever have one without rewriting the laws of physics. Examples include psychic phenomena, accounts of ghosts and other-worldly apparitions, alien abductions or fairies at the bottom of the garden, all of which not only fall outside of mainstream science, but which have no basis in reality.

It is this third category that many people understandably find the most fascinating; in fact, the stranger the mystery, the better. Of course, none should be taken too seriously, since they can all be explained away rationally, but where is the fun in that? Mysteries of the Third Kind are not true mysteries; they are fictions—stories we have shared with each other across cultures and across the ages. Some of them may at one time have been considered Mysteries of the Second Kind, when there might have been hope of a rational explanation. Yet they remain important to us—even after we've learned that they are not true—as mythologies, folklore and fairy tales, and of course fodder for Hollywood movies, for without them our lives would be poorer.

It is when Mysteries of the Third Kind cross over from harmless beliefs (like the existence of ghosts, fairies, angels or alien visitors) into dangerous irrationalities that they can have detrimental effects on our wellbeing, such as when those claiming to have psychic powers dupe the innocent and vulnerable, or when those peddling alternative therapies and quack remedies denounce established medical treatments or refuse their children vital vaccines. That is when we can no longer stand by and do nothing.

What I want to focus on in this lesson is Mysteries of the Second Kind—real mysteries that we are still searching for answers to. One of the most surprising things at the heart of science is that the laws of nature are logical and comprehensible. But it need not have been so. Before the birth of modern science, our beliefs were ruled by myths and superstition (Mysteries of the First Kind)—the world was unfathomable and inexplicable, known only to a divine higher power. We were content with the mysteries we encountered, and even celebrated our ignorance. But modern science has shown us that by being curious about the world—by asking questions and making observations—we discover that what was once a mystery can be understood and explained rationally.

Some people argue that the cold rationalism of science leaves no room for romance or mystery. Nervous about the rapid advance of science, they feel that the act of searching for answers to things we do not yet understand somehow detracts from their awe and wonder. One reason for this view may be because modern science has shown the universe to have no purpose or end goal, and that humans have evolved on Earth through the process of natural selection based on random genetic mutations and survival of the fittest. This is seen as too bleak an explanation for our existence and implies that our lives have no meaning. When I have found myself explaining my work to non-scientists at social gatherings or dinner parties I have sometimes been made to feel like the "learn'd astronomer" in Walt Whitman's poem[15]—a killjoy, ruining the magic and romance of the stars with tiresome logic and rationalism. But to think that way is misplaced. Many scientists like to quote the American physicist Richard Feynman, who was frustrated by an artist friend of his who could not appreciate what science can give us:

Poets say science takes away from the beauty of stars—mere globs of gas atoms. Nothing is

15 Walt Whitman, "When I Heard the Learn'd Astronomer" (1867), https://www.poetryfoundation.org/poems/45479/when-i-heard-the-learnd-astronomer.

'mere'. I too see the stars on a desert night and feel them. But do I see less or more? . . . What is the pattern, or the meaning, or the why? It does not do harm to the mystery to know a little more about it. For far more marvellous is the truth than any artists of the past imagined it. Why do the poets of the present not speak of it?

Unlocking nature's secrets requires inspiration and creativity no less impressive than anything in art, music or literature. The sense of wonder at the nature of reality that science continues to reveal is the polar opposite of the dry, hard facts that some imagine science to be.

You might be surprised to know that many particle physicists were secretly hoping that the famous Higgs particle, discovered at the Large Hadron Collider in 2012, would in fact *not* be found—that despite our best mathematical theories of the fundamental constituents of matter predicting its existence, and despite the years of effort and billions of dollars spent building one of the most ambitious science facilities the world has ever known to hunt for it, it would have been even more exciting if we had confirmed its nonexistence!

You see, if the Higgs *didn't* exist, then this would have meant that there was a flaw in our understanding of the fundamental nature of matter and that we would have needed to find a different explanation for the properties of elementary particles—an exciting new mystery to solve. Instead, the discovery confirmed what we already suspected. To the curious scientist, verifying an anticipated prediction is less thrilling than a truly unexpected discovery. Now, I don't want to give you the impression that physicists were unhappy about the confirmation of the Higgs. We still celebrated its discovery, for learning more about the universe, whether the result is a surprise or not, is always better than remaining in ignorance.

Striving to understand the world around us is a defining feature of our species, and science has given us a means to achieve this. But it allows us to do more than simply solve scientific mysteries just for the sake of it; it has also ensured the survival of the human race. Let us go back to the fourteenth century—a time before modern science—and consider the horrific devastation of the Plague (also known as the Black Death), which, together with the Great Famine that had taken place a few decades earlier, killed up to half the population of Europe.

Apart from the terrible loss of human life, the Plague had huge social consequences. Without the benefits of a modern scientific understanding of the disease (or the bacterium, *Yersinia pestis,* that caused it)—let alone antibiotic medications to treat the sick—many people turned instead to religious fanaticism and superstition. And since no amount of prayer seemed to help, they believed that the epidemic therefore had to be a punishment by God for their sins. Many reacted in horrific ways in an attempt to gain God's forgiveness, for example by scapegoating and killing those they perceived as heretics, sinners, and outsiders: Romani, Jews, friars, women, pilgrims, lepers and beggars—it didn't matter. But remember, this was the medieval world, when most occurrences were attributed to magic or the supernatural. You might argue that they simply didn't know any better.

Fast-forward seven centuries to the modern world and the way humanity has dealt with the Covid-19 pandemic. Science has enabled us to understand the coronavirus that caused it, and scientists quickly mapped its genetic code in detail. This allowed for a range of vaccines to be developed, each delivering in its own clever way the genetic instructions to the cells in our bodies to manufacture the molecular ammunition (the antibodies) to protect us against the virus if it ever did attack us. Today, diseases are no longer a mystery. Most of us don't have a deep knowledge of the nature of the coronavirus or the disease it causes and how that spreads. However, we are grateful to those who have solved this mystery, and it is a sad indictment of the modern world that there are still many who prefer to reject this knowledge, even while arguing that they are themselves being rational and enlightened.

Nothing expresses more clearly the importance of curiosity about the world and the value of enlightenment over ignorance than Plato's

Allegory of the Cave. It goes something like this. A group of prisoners have lived all of their lives chained to the floor of a cave, facing one of its walls such that they are unable to turn their bodies or heads around. Behind them and unbeknownst to them there is a burning fire and a steady stream of people passing in front of it, thus casting shadows on the wall that the prisoners are facing. For the prisoners, these shadows represent their entire reality, for they are unable to see the real people behind them who are making the shadows. The sounds of the people talking are heard by the prisoners who, because of the echoes around the cave, are fooled into thinking they are coming from the shadows themselves.

One day, one of the prisoners is set free. When he steps outside the cave he is first blinded by the bright sunlight and it takes him time to adjust. Eventually, he begins to see the world as it really is, with three-dimensional objects and light reflecting off of them. He learns that shadows are not objects in themselves but only formed when solid objects block the passage of light. He also learns that this outside world is superior to the one he had experienced inside the cave.

When given the opportunity, he goes back inside the cave to share his experience with the other prisoners, whom he pities for not having experienced true reality, only their limited one. But the prisoners think that their returning friend is insane and refuse to believe him. And indeed, why would they? The shadows they see are all they have ever known, and they cannot comprehend another version of reality, so they have no reason to be curious about the origin of the shadows or how they are formed by the interplay between light and solid matter. Can one argue that their reality, their truth, is as valid as his? Of course not.

According to Plato, the prisoners' chains represent ignorance, and we cannot blame them for accepting their limited reality at face value based on the evidence and experiences they have, but we also know that there is a deeper truth. Their chains stop them from being able to seek this truth.

In the real world, our chains are not so restrictive, for we *can* be curious about the world and we can ask questions. Like the freed prisoner, we know that whatever reality we are experiencing, we may still have a limited perspective. We are viewing reality from one frame of reference. In other words, even the freed prisoner might reflect on the possibility that he has simply stepped into a larger 'cave', which is still not showing him the 'complete' picture. Likewise, we should acknowledge that our view of reality might also be limited, since mysteries still exist. However, we should not be content with accepting these mysteries but should always try to gain a deeper understanding.

Although Plato's allegory of the cave goes back over two millennia, there are modern versions of it, particularly as depicted in a number of Hollywood movies, such as *The Truman Show* and *The Matrix*. In both these films, a curiosity about the nature of reality leads to enlightenment—to seeing things as they really are. Whether that is itself the ultimate reality or not, it is still a step closer to the truth and is thus always preferable to remaining in ignorance.

My point is that science does not try to dismiss mysteries, as some people might claim. In fact, quite the opposite: it acknowledges that the world is full of mysteries and puzzles that it then tries to understand and solve. If there is strong scientific evidence that an unexplained phenomenon is real, and yet it does *not* fit into the existing body of knowledge, then that is the most exciting outcome of all, for it points to new discoveries and new knowledge to be gained. Put another way, the enjoyment we derive from doing a jigsaw puzzle is the process of connecting the pieces together. Once it is finished there is a short-lived sense of satisfaction in being able to admire the complete picture, but that doesn't last very long. In fact, if we're keen on jigsaws, we'd already be looking forward to starting a new one. This should apply in everyday life too. There are many mysteries out there, but their true appeal is in trying to solve them, not in leaving them be.

We all constantly encounter things in life that we do not understand, that are new or unexpected. This is not something to lament or fear. Encounters with the unknown are normal and you don't need to shy away from them. At the heart of science is curiosity—to question, to want to know. We are all born scientists; as children we learn to make sense of our world by exploring and asking questions all the time. Thinking scientifically is in our DNA. Why then do so many of us stop being curious about the world when we reach adulthood, becoming complacent, content even, about that which we don't understand?

It need not be this way. We should all ask questions when faced with a mystery, to free ourselves from the 'chains' of ignorance and look around us. Ask yourself whether you are seeing the whole picture and how you might find out more.

I am not suggesting, of course, that everyone has to always be on the lookout for things to understand and explain; after all, some people are just less curious than others—and daily life might get a bit intractable if we all behaved in the same way, going around poking our noses into everything, tilting at windmills, not accepting stuff we don't understand even when we know there are people who do understand it, feeling compelled to reinvent the proverbial wheel over and over. In any case, most people don't have the luxury of time or resources to go around solving mysteries all the time, even if they wanted to. If you fall into this category, then what is the value of this lesson to you? If you are ever faced with the inexplicable or the bizarre, then of course, it can often be more fulfilling to simply enjoy the mystery—like a fun or perplexing conjuring trick that would be spoiled if we knew how it was done—and that is just fine. But be aware that there are plenty of other examples in daily life that would give you greater joy and fulfilment if you were able to understand them. Enlightenment is almost always preferable to ignorance. If you're unshackled from your chains, take that opportunity to step outside of the cave and into the light of the Sun.

4

IF YOU DON'T UNDERSTAND SOMETHING, IT DOESN'T MEAN YOU CAN'T IF YOU TRY

Just as we come in all shapes and sizes physically, so too do our brains differ in the way they function. But we should not use this as an excuse not to try to understand something. Almost nothing is beyond your understanding if you put your mind to it. Remember that *anyone* with a deep knowledge of a subject—whether they're a plumber or a musician, a historian or a linguist, a mathematician or a neuroscientist—will have gained this knowledge through dedication, time and effort.

I am not saying that we all have equal mental capacity to comprehend difficult concepts. Just as there are people who are born athletes or who are gifted musically or artistically, so there are those who are mathematically minded or naturally good at logical thinking. Likewise, some of us have good memories—if you are not one of these people then no doubt you will know friends or family members who are; they are the ones who always do well in quizzes because they are able to retain and retrieve a lot of information. I'm not one of those people, which is why I preferred physics over chemistry and biology at school, as it didn't require me to remember as much 'stuff' (or so I thought of these subjects at the time).

Many of us have at some point in our lives experienced what is known as imposter syndrome—the feeling that we are not up to a task entrusted to us or that others' expectations of our ability are higher than our own. This often manifests itself when we start a new job and are surrounded by people at ease with what they are doing and who seem to know so much more than we do.

We feel justified in having this sense of doubt and insecurity because, we tell ourselves, we know our abilities and capabilities better than anyone else. We are convinced we are not good enough and worry that soon everyone else will realise this too and our cover will be blown. This is a perfectly natural reaction to being exposed to something new that we need time to become familiar with.

Nowhere is this more common than in science. The regular research seminars in my physics department at the University of Surrey are delivered to a mixed audience ranging from PhD students to senior professors. Unless they are particularly self-assured, most of the students tend not to feel confident enough to interrupt a speaker to ask for clarification on something they've said, for they imagine that their shallow understanding of the subject matter will be exposed. What I find amusing is that, more often than not, it is the senior professors who will ask the 'dumbest' questions. Sometimes, this is because what at first appears to be a really basic question turns out instead to be deeply insightful. But, more often than not, it isn't. Here is my point. Only someone very familiar with the subject matter of the seminar would perceive it as a basic question. Professors are well aware that they cannot be expected to know everything, especially if the topic is outside of their area of expertise, so there is no shame in exposing their ignorance. They may also wish to ask the question on behalf of others in the room, like their students, who might not have the confidence to do so themselves.

When it comes to wider society, the reason scientists like me try hard to communicate scientific ideas is in part that we see the value of a scientifically literate populace. Whether it is to play a part in controlling a global pandemic, tackling climate change, protecting the environment or adopting new technologies, it helps if wider society has some level of understanding of the underlying science, which requires not only the effort to learn a little about the issue, but the willingness to do so. We have seen this clearly during the pandemic, in which the public has been asked to 'trust the science' and to 'follow the scientific advice' on social distancing, wearing face masks, and acting responsibly in a variety of ways.

A lot of people I meet are intimidated by complex ideas that are unfamiliar to them. If I try talking to them about some topic in science—maybe something I am working on in my research—they will shy away from engaging with me. It may well be that they would simply prefer to turn the subject to a more interesting topic (to them). However, if they are expressing a lack of confidence in being able to understand and engage with science, then I want to address that head-on, because this attitude can be deeply detrimental, and contagious; worse than that, they might pass this attitude on to their children, putting them off science too, along with all the good mental habits that the scientific method teaches us. And that would be downright tragic.

One of the lessons a scientist learns early on is that whenever there is a concept he or she doesn't understand, it is most likely because he or she hasn't had the benefit of the time and effort needed to study it. I am a physicist, which means that I feel confident talking about the nature of matter, space, time, and the forces and energies holding the universe together at a fundamental level. But I have little knowledge of psychology, geology or genetics; I am just as ignorant of these areas of science (and others) as anyone else. However, this does not mean that with dedication and sufficient time I couldn't eventually become an expert in them. This is not hubris, because by 'sufficient time' here I mean years, and probably decades, of study, not hours or days. However, I can still have an interesting and informative conversation with experts in these fields, provided they don't get too technical and I give them my full attention. This is something I have been doing over the past decade that I have been presenting my BBC Radio 4 programme, *The Life Scientific*, in which I discuss a wide range of scientific subjects with leaders in their fields. I don't have to be an expert myself, just sufficiently interested and curious, neither of which requires scientific training. This is generally true in other walks of life, too.

I am not suggesting that every one of us needs to train as an epidemiologist or virologist in order to protect ourselves during a pandemic. No one— not even the brightest physicist or engineer—

understands all the technology that goes into, say, a modern smartphone; nor does any one person need to, certainly not to be able to make full use of one. Knowing how to use the apps on the phone does not require a deep understanding of how all the electronic components inside it work. There are other situations in life, however, where it can be beneficial to have more than a superficial grasp of a subject—because it helps us make important decisions, such as understanding the distinction between bacterial and viral infections and the fact that only the former can be treated with antibiotics, while vaccines help people avoid the latter.

At this point I feel I should give an example of what I mean by a difficult concept in science, one which you may think is beyond your ability to comprehend. Please humour me by reading through the next few pages. If you are able to follow, then it is to your credit alone and not to my skills as an explainer, for it is far easier to explain something one knows well than it is to comprehend a new and difficult concept.

Consider the following puzzle. If you were able to fly at the speed of light while holding a mirror in front of your face, would you see your own reflection? After all, to see your reflection in a mirror requires light to leave your face, reach the mirror in front of you, and then reflect back into your eyes. Since we are confident that nothing, according to the laws of physics, can move faster than the speed of light (recall the infamous story of the faster-than-light neutrino experiment), if you are moving *at* the same speed as the light, how can it move away from your face to reach the mirror, which is itself moving away from the light at the same speed? Surely you would not be able to see your reflection, just like the mythical vampire. Well, you would be wrong to think this. How can this be? Let's solve this puzzle together.

Imagine that you are on a train and another passenger walks past your seat in the direction of the train's travel. Since you and she are both moving with the train, she walks past you at the same walking pace she would do if the train were stationary. However, at that very moment the train passes through a station without stopping and someone on the platform also sees the passenger walking through the train. To him, the passenger is moving past at a speed that is the combination of her walking pace *and* the much faster speed of the train itself. So, the question is this: how fast is the passenger *really* moving—at the walking pace that you measure (sitting there on the train), or at that speed *plus* the train's speed, as measured by the outside observer? If you think the speed of the walker as measured by the person on the platform is their 'true' speed, then think about the fact that the train is speeding along a track on Earth, which is spinning around on its axis and also moving along its orbit around the Sun. Maybe to someone floating out in space, the train looks stationary while the Earth moves beneath it. The answer to the question of how fast the passenger is really moving is that you on the train and the observer on the platform are *both* right, in your own frames of reference, for there is no single true value for the speed of the walker. All motion is relative. This is an idea at the heart of the aptly named theory of relativity.

Now let us turn to the nature of the speed of light. We learn at school that light is a type of wave, and waves need something to travel through; some 'stuff' that does the 'waving' or vibrating. For example, sound waves moving through the air need the air to travel through because sound is nothing more than vibrations of the air molecules themselves. This is why there is no sound in the vacuum of space. It stands to reason then that light waves should also need something to travel through, and nineteenth-century scientists set out to find out what that something might be. After all, unlike sound waves, light reaches us from distant stars by travelling through the vacuum of space. It was therefore assumed that there must exist an invisible medium—called the 'aether'—filling all of space and carrying light waves through it. Scientists designed a famous experiment to test for its existence, but found no evidence of it. It was Einstein who showed that light always moves through space at the same speed regardless of how fast we ourselves are moving when we measure it. Going back to our example of the train, it's as if you (on the train) and the observer on the platform both measure the speed of the person walking through the train to be the same. How can this be possible? It sounds crazy, but it turns out that this is indeed how light behaves.

Now for the next step. Consider two astronauts onboard spaceships approaching each other at high speed in empty space. Since all motion is relative, the astronauts cannot determine if or how fast each is moving individually, only that their ships are getting closer to each other. One of the astronauts shines a light beam towards the other and measures the speed of the light as it leaves him. (If we compare this to our train example, the speed of the light beam is like the speed of the person walking on the moving train.) Since this astronaut can quite legitimately claim to be stationary, with the other ship doing all the moving, he should see the light moving away from him at a billion kilometres per hour (the measured speed of light, which we now know very well). At the same time, the other astronaut can also legitimately claim to be stationary (from her perspective, the other ship could be the one that is moving), and she too measures the speed of the light reaching her to be the same one billion kilometres per hour—no more, no less. So, both measure the light to have the same speed, even though they are clearly moving relative to each other!

Incredible though this may sound, we at least have the answer to the puzzle I posed earlier. Flying through space at the speed of light with a mirror in front of your face, you will indeed still see your reflection because, regardless of your speed, the light will still leave your face at a billion kilometres per hour, hit the mirror and reflect back into your eyes, just the same as if you were not moving at all. The speed of light in a vacuum is a fundamental constant of nature; it has the same value, no matter how fast an observer is moving. This is one of the most profound ideas in science, and it took no less a genius than Albert Einstein to figure it out.

Following Einstein's argument in detail would require more explaining than we need to go into here and now, but it can be understood by anyone prepared to invest the time and effort.[16]

---

16 There are many books that explain Einstein's ideas in simple terms without expecting the reader to have a back-

We're all capable of digesting more-complicated ideas than we may initially give ourselves credit for. Some ideas and concepts take time and effort to grasp, and that is ok. Even if we're not all as smart as Einstein, even if we do not have as much training in physics and mathematics, with an open mind and some effort we can still gain an understanding of some of the concepts at the heart of his ideas and equations.

We don't all need to be Einsteins or even physicists to appreciate how light behaves or to understand something profound about the nature of space and time, in just the same way we don't need to have studied vaccinology to understand that getting a flu jab will protect us. We can stand on the shoulders of giants, lean on the strengths and knowledge of others who have put in the years to gain expertise that can then be shared with the rest of us. So, even if we encounter something we don't understand

---

ground in physics. All you need is to want to find out more. For example, I say more about the nature of light in my book *The World According to Physics*.

right away, we can still make an effort and take some time to try. Sometimes it is for no better reason than to expand our minds; sometimes it can help us make a decision that will benefit us in our daily lives. Either way, we are the richer for it.

Of course, one of the features of modern life, mostly thanks to the internet, is that we all have to constantly make choices about what to pay attention to—what to spend our time on, even if it is for just a few minutes. Many of us today have instant access to far more information than we can ever hope to process, which has meant that our average attention span is getting shorter. The more 'stuff' we have to think about and focus on, the less time we are able to devote to each particular thing. People are quick to blame the internet for this reduced attention span, but while social media certainly plays its part, it is not entirely to blame. This trend can be traced back to when our world first started to become connected early in the last century as technology gave us access to an ever-increasing amount of information.

Today we are exposed to twenty-four-hour breaking news and an exponential rise in the amount of produced and consumed information. As the number of different issues that form our collective public discourse continues to increase, the amount of time and attention we are able to devote to each one inevitably gets compressed. It isn't that our total engagement with all this information is any less, but rather that as the information competing for our attention becomes denser our attention gets spread more thinly, with the result that public debate becomes increasingly fragmented and superficial. The more quickly we switch between topics, the more quickly we lose interest in the previous one. We then find ourselves increasingly engaging only with those subjects that interest us, leading us to become less broadly informed—and potentially less confident in evaluating information outside of the spheres with which we are most familiar.

I am not advocating that we should all devote more time and attention to every topic we encounter, whether we are exposed to information through our family, friends or work colleagues, or by reading books and magazines, the mainstream media, online or on social media, as that would be impossible. But we must learn how to discriminate between what is important, useful and interesting, what is deserving of our attention and time, and what is not. As Feynman so emphatically pointed out, in his response to the journalist's request for a pithy summary of his Nobel Prize work, the topics we do choose to spend more time thinking about and digesting will inevitably require a certain level of commitment. In science, we know that to truly understand a subject requires time and effort. The reward is that concepts which may at first have seemed impenetrable turn out to be comprehensible, straightforward, sometimes even simple. At worst, we acknowledge that they are indeed complicated—not because we are unable to think them through thoroughly and make sense of them, but because they just *are* complicated.

So, this is the takeaway for us all in daily life. Do you need a PhD in climate science to know that recycling your rubbish is better for the planet than throwing it all in the ocean?

Of course not. But taking some time to dig a little deeper into a subject and weighing up the evidence, the pros and cons about an issue, before making up your mind can help you make better decisions in the long run.

Most things in life are difficult to begin with. But, if you're prepared to try, you can cope with far more than you imagine.

5

DON'T VALUE OPINION OVER EVIDENCE

A few weeks ago, my plumber came round to fix our boiler, which had been switching itself off intermittently. I informed him that I had seen an error message appearing on the boiler's display that just said 'F61'. He said he knew what this meant and that a circuit board probably needed replacing. He told me that this would fix the problem because he had dealt with hundreds of boilers with the same issue and his solution always fixed them. I trusted his judgement and I was right to do so, because the boiler is now working fine. There is no way I would have had any idea how to fix the boiler myself, but I trust my plumber because he is an expert. I also trust my dentist, my doctor and the pilot who flies the plane I'm on.

But how do we decide who, or what, can be trusted? The reason I believe we need to break this down is because, as we encounter information on a daily basis, we need to decide what is valid and legitimate—backed up by facts and reliable evidence, for example—versus what is mere opinion. This is becoming increasingly important when so many of the decisions we make every day, both as individuals and collectively as a global community, need to be based on critical analysis and trustworthy evidence.

These days, a great many people regard themselves as experts with a licence to speak with perceived authority on all manner of topics, often based on nothing more than an inflated sense of their own sagacity. The reason for this seems clear to me: our easy access to the internet has democratised information to such an extent that some feel empowered not only to hold ill-informed or unsavoury views, but to inflict them on others with the assurance and confidence that was once the preserve of preachers and politicians. This does not necessarily mean that they are wrong, of course. So, how can we be sure that what we are told, or what we read, can be believed? How do we disentangle evidence-based established facts from uninformed opinion?

Tragic though it has been and continues to be for countless millions of people around the world, the coronavirus pandemic has highlighted better than any event in modern times just how important it is to heed scientific advice that is based on reliable evidence. But we need to know what constitutes trustworthy and reliable evidence, and that is not as straightforward as you might think.

Some will say that they know good evidence when they see it. But that is not enough. As humans, we may sometimes only see what we want to see or what we expect to see. If that happens, confirmation bias sets in (see the next chapter), and we end up putting our trust in the flimsiest of evidence so long as it supports what we already think. No, sound evidence needs to be objective, unbiased, and built on solid and reliable foundations. It needs to come from a trusted source or sources and to be free of inconsistencies and alternative interpretations. If you have ever sat on a jury and been asked to reach a decision in a court case, then you will have had to think critically, objectively and, as best as you can, without bias. In short, you will have had to think scientifically.

One of a number of definitions of 'science' is that it is *the process of formulating meaningful statements, the truth of which is only verified through observational evidence*. And as a way of distinguishing between scientific knowledge and other belief systems, such as religion, political ideologies, superstitions or even subjective moral codes, which do not require supporting evidence or verification in the same way, this definition is a strong start. But its weakness is that it doesn't tell us how *much* evidence we need and of what *quality* it must be. This is called 'the problem of induction'.

Of course, the more evidence we can amass, the more reliable our knowledge becomes, so who gets to decide what is reliable evidence and what isn't? And how can we tell if there is enough evidence for us to feel confident that something is true? Well, it depends on what we want to use that evidence for and the potential cost of making the wrong decision based on using it. Even a tiny amount of evidence suggesting there is a harmful side effect of a new drug should be enough to immediately stop its use until the matter can be better understood, whereas we should demand a lot of evidence to convince us of the existence of a new subatomic particle.[17]

Related to the problem of induction is something known as 'the precautionary principle'. Basically, what do we do if the evidence is poor or incomplete? Here we must weigh up the cost of trusting the evidence, and possibly acting on it, against the cost of inaction. Many climate change

---

17 "'Extraordinary claims require extraordinary evidence' was a phrase made popular by Carl Sagan who reworded Laplace's principle, which says that 'the weight of evidence for an extraordinary claim must be proportioned to its strangeness'". Patrizio E. Tressoldi, "Extraordinary claims require extraordinary evidence: the case of non-local perception, a classical and Bayesian review of evidences," *Frontiers in Psychology* 2 (2011): 117, https://www.frontiersin.org/articles/10.3389/fpsyg.2011.00117/full

sceptics argue that scientists cannot be certain that anthropogenic climate change ('anthropogenic' meaning 'the result of human activity') is taking place. This is true: they can't be certain because nothing is 100 percent certain in science (though as I've said before, that doesn't mean that there aren't established 'facts' about the world). But there is overwhelming evidence pointing to humankind as being responsible for the way Earth's climate has been changing so rapidly over the past few decades; and erring on the side of caution is in any case preferable to ignoring the evidence and doing nothing. Imagine your doctor telling you that you only have just a few years left to live unless you change your lifestyle in some way—for example, by giving up alcohol and smoking. She tells you that while she cannot be certain that making the change will achieve the desired outcome, she is nevertheless 97 percent sure she is right.[18] Would you say, "Well, Doc, if you're not

18 According to many surveys, roughly 97% of climatologists believe that humans are dramatically and adversely affecting Earth's climate.

completely sure, then there's a chance you are wrong, so I am going to carry on doing what I'm doing because I enjoy it"? Chances are, even if the doctor claimed she was only 50 percent sure, you would probably still try to heed her advice, wouldn't you? Maybe not. Maybe it would be too hard for you to change your lifestyle, or maybe you are prepared to take a gamble.

The precautionary principle comes with caveats, however. When politicians have to make important policy decisions that affect the whole of society, the scientific evidence, however compelling, may not be the only consideration. We have seen this during the pandemic, with tighter restrictions to slow the spread of the virus coming at the expense of damage to economies, the loss of livelihoods, and the impact on the mental health and well-being of many vulnerable people. Sometimes, despite strong scientific evidence in support of a particular course of action, it has to be seen as part of a wider, more complex issue— and, of course, as individuals, we will all have different circumstances that need to be take into account as well.

Another problem is that this requirement for supporting evidence can become muddled when one hears a scientist saying that they 'believe' something to be true. Scientific 'belief' does not have the same meaning as when the word is used more casually in everyday language; it is not, or at least shouldn't be, based on ideology, wishful thinking or blind faith, but rather on tried and tested scientific ideas, observational evidence and past experience built up over time. When I say that I 'believe' the Darwinian theory of evolution to be true, I am basing that belief on the mountains of available evidence that support evolution (and the lack of credible scientific evidence that could disprove it). While I have not trained as an evolutionary biologist myself, I trust the expertise and knowledge of those who have, and I feel I am able to tell the difference between strong evidence based on a lot of good science and mere opinion that is based on blind faith, prejudice or hearsay.

Scientists, like any experts in their field, can of course get things wrong, and no one is expected to trust them blindly or unconditionally;

rather, one should check to see if what they are saying is accepted by others. However, this doesn't mean you can shop around until you find an opinion you like or one that supports your preconceived views. If I have a health concern, I may be able to learn more about it by spending an evening doing online research, so that I can ask better questions about my treatment options the next time I talk to my doctor; but I would not argue with someone who has far more expertise and experience than I do on some subject solely because their opinion doesn't appeal to me.

Like any experts, scientists can be trusted to know what they are talking about, not because they are special, but because they have devoted years to studying and building up that expertise. I am an expert in quantum physics, but that does not give me any special insights into plumbing, playing the violin, or flying a plane, although I might well have been able to do any of those things competently if I had spent years doing the necessary training. I won't argue with my plumber over how to fix my boiler, and he won't tell me about how to diagonalise a Hamiltonian.[19] That said, questions are always welcome. And, in return, what you should expect and demand is expertise and evidence, not unfounded opinion.

Of course, simply claiming expertise in a subject is not enough. A ufologist who has spent years examining evidence for the existence of aliens might be considered an expert too. Similarly, conspiracy theorists who think the Earth is flat would argue passionately that there is ample evidence in support of their claim and that it therefore satisfies the test of verification, and hence must be true. Should we reject their views because they don't hold a PhD or because they don't belong to an exclusive science 'club'? Of course not. But while it is important to remain open-minded to new ideas and other points of view, we shouldn't be so open-minded that our brains fall out. A healthy level of open-mindedness goes hand in hand with scrutiny and critical enquiry.

19 A technical reference to a mathematical technique in theoretical physics called matrix mechanics.

We all know someone who subscribes to a particular conspiracy theory, whether they are driven by political ideology or were just innocently watching YouTube videos and got sucked in. But conspiracy theories are as old as human civilization itself; for as long as the powerless and disenchanted have resented being kept in the dark, they have speculated about matters they do not comprehend. Though they may truly have been lied to and deceived, it is just as likely that their theories are completely unfounded. And this is not to say that anyone who believes a particular conspiracy theory is simply not smart enough to see through it. Many intelligent and otherwise well-informed people may have valid reasons for believing in something that is not true, whether it's because of a legitimate distrust of authority based on some past experience or simply because they don't have access to all the facts. In this case, it does no good telling them they are wrong because they are not clever enough to see the truth. They will feel exactly the same way about you.

Ask yourself this, however: When was the last time a conspiracy theorist actually uncovered a real conspiracy? When have they ever proven, beyond doubt, that they were right? If you think about it, that's the last thing a conspiracy theorist actually wants—because the conspiracy itself is their raison d'être. Their mission to uncover the 'truth' is what drives them and comforts them. It defines who they are. Conspiracy theorists are sustained as much by the passions that their arguments instil in them as by the rational arguments that they try to make to support their claims. And just as they never succeed in uncovering their conspiracy, they are at the same time utterly unshakeable in their belief that they are right. The notion that the underlying premise of their theory is unfounded is never entertained. Ask a conspiracy theorist what evidence it would take to change their minds— and they will have to admit that nothing would. In fact, when presented with evidence against their theory, they see it as merely confirming the lengths to which those they believe are behind the conspiracy will go in order to keep the truth from coming out. A conspiracy theory is by its nature irrefutable.

How different this is from the way we do science, where we try our utmost to *disprove* a theory, for only in that way will we build confidence that our understanding of the true nature of reality is robust—and potentially uncover something new about the world.

The reason I have focussed on the distinction between scientific theories and conspiracy theories is because this can help us appreciate the various types of evidence that can be put forward to back up a claim. This is more important today than ever before because of the speed with which certain ideas can spread on social media. When someone believes the Earth is flat, the Moon landings were fake, or, more exotically, that aliens have visited Earth—whether the US government is covering up evidence of an alien spacecraft crash site in Roswell, or aliens are behind the building of the Pyramids of Giza— this can be seen as benign and innocuous, amusing even. But when we hear of conspiracy theories claiming that Covid-19 is a hoax and part of a government strategy to control us, or that all vaccines are harmful or (again) part of a strategy to control us, then this can no longer be ignored or dismissed as harmless fun. We need to be able to assess such claims objectively and scientifically.

Tackling conspiracy theories is being taken more seriously now than ever before, with social media platforms trying harder to weed out misinformation and fake news. But there is plenty we can do as individuals to empower ourselves in the meantime. Firstly, we can all be more mindful of the problem and take measures to combat it. It is worth remembering that most people who subscribe to conspiracy theories are perfectly well-adjusted and sensible individuals who have been taken in by those who feed on fear, insecurity, and feelings of disenfranchisement, particularly during a time of crisis, which can be a very effective time in which to sow seeds of doubt and fan the flames of all manner of false ideas.

Applying a scientific approach when evaluating a particular idea, claim or opinion—whether it is posted by a friend on Facebook or whether it comes up in conversation—can often help you separate truths from untruths or to reveal contradictions within the idea. Therefore, try to look beyond the superficial claim, ask questions, and examine the quality of the evidence supporting it. Ask yourself how likely the claim is to be true and whether those advocating it have motives for doing so: Are they being entirely objective, or do they have an ideologically driven reason for holding these views? Challenge the evidence: Where did it originate from, and is the source credible? And remember that even the most outlandish conspiracy theory can be built on a kernel of truth. The problem is that such a truth can be used to power and sustain an ever-growing and preposterous edifice around it, made up of half-truths, unsupported claims and outright falsehoods.

It can often feel frustrating and pointless to argue with a conspiracist. Highlighting logical contradictions or a lack of reliable evidence, even showing them evidence that counters their claim, can seem like a waste of time when you don't make headway in changing a person's mind. But that does not mean you shouldn't try. What you should not to do is accuse another person of ignorance or stupidity, however tempting that might be in the heat of an argument. Instead, examine where they have obtained their evidence; ask them what the chances are that the conspiracy could have been kept a secret by so many people. The Moon landing hoax is a good example of a conspiracy theory that is difficult to justify on these grounds, as it requires tens of thousands of people working for NASA and the many other industries that supported the Apollo missions to have been 'in on it' and to have remained silent for half a century. Equally important, try to understand their underlying concerns and *why* it is that they believe, or want to believe, what they do.

We cannot all be expected to crusade against every opinion or belief we disagree with, but what we can do instead is to take these opportunities to assess our own beliefs. Remember that the scientific method is a process of critical thinking, questioning and holding up theories—both ours and those of others—to the light of empirical evidence. This is how we test and verify ideas about the world. It is an approach we should all try to adopt in everyday life. While we should always question the opinions and beliefs of others and reflect carefully on whether they are based on reliable evidence, ultimately what matters is what *we* believe and why.

So, ask yourself why you hold the views you do and, in turn, whose opinions you trust and why. Ask yourself whether you would rather trust someone who expects you to blindly accept what they tell you without question or who gets angry or tries to shut you down if you don't—or someone whose entire life philosophy is based on asking questions and seeking answers, even if those answers lead to a profound change in what they thought. Ask yourself if you think others should trust what you say—and why. And remember, evidence will prompt questions. So, value questions: ask them and encourage others to do so— and always demand that the answers (both those you receive and those you give) are, like the questions, thoroughly grounded in evidence.

6

RECOGNIZE YOUR OWN BIASES BEFORE JUDGING THE VIEWS OF OTHERS

We all tend to feel more comfortable protected within our bubbles, surrounded by like-minded people. That is only human nature. But these bubbles are also echo chambers where we expose ourselves only to opinions and beliefs we already agree with. Within these echo chambers, our views get amplified and reinforced through repetition and confirmation, and so we build up prejudices and preconceptions that are then very difficult to shake. Consciously or unconsciously, we succumb to what is known as confirmation bias. It is a hard fact of life that we are often capable of recognising the biases in others' views while hardly ever questioning our own beliefs. Being a scientist does not make one immune to confirmation bias. But thinking scientifically can help inoculate us against it and other blind spots, and that's the message I want to share in this chapter. Let me give you an example.

I have little doubt that the climate is changing rapidly and that this is due to humankind's actions; and I am confident that if we don't all work together to change the way we live our lives, then the future of humanity is in peril. I base my view on overwhelming and incontrovertible scientific evidence that comes from many different areas of science: climate data, oceanography, atmospheric science, biodiversity, computer modelling and so on. Imagine being dissatisfied with the prognosis of your doctor and not just seeking a second opinion from another qualified physician, but getting a third, fourth and fifth opinion, with all of them telling you the same thing, all backed up by irrefutable supporting evidence, such as blood tests, scans and X-rays. That is why I hold the view I do about climate change.

But maybe I should reflect on why I was so ready to accept the 'truth' of anthropogenic climate change years ago, before the mountains of supporting evidence built up. Was it because I knew several climatologists personally and trusted their expertise—maybe because I regarded them as honest and competent scientists? Or did it also have something to do with my liberal views about leading a more ethical life and protecting the world's natural resources, together with my opposition to libertarian views that, as I see it, value personal freedoms over sustainable living? You can see that, even as I write this, my personal biases are clear, and it is very hard to remain completely objective on the matter. As it happens, when it comes to anthropogenic climate change, the scientific evidence is now so strongly weighted in favour of it being true that I don't have to question my motives for having 'believed' it from the start. Nevertheless, I have no doubt that, however objective I try to be, I was probably more prepared to accept evidence in support of anthropogenic climate change and more cynical and distrusting of any evidence purporting to show that it wasn't taking place. This is confirmation bias at work.

Confirmation bias comes in many forms, and psychologists have identified various ways it can manifest itself. One is the phenomenon of illusory superiority, whereby a person will have an overinflated sense of their own competence while at the same time failing to recognise their weaknesses. There have been a number of studies over the past few decades on why people can be so unaware of their incompetence or shortcomings in understanding concepts and situations. This can sometimes be amusing, such as the case of the bank robber, McArthur Wheeler, who believed that he could hide his face from security cameras if he covered it in lemon juice; he had misunderstood the chemistry behind the way invisible ink works. But when people suffering from illusory superiority are in positions of power or influence, this can have dangerous consequences if they gain large numbers of followers who are taken in by them. For example, we know that those who shout the loudest on social media often have the largest numbers of followers and yet often tend to be the ones most likely to be suffering from illusory superiority.

Much of the research into illusory superiority was carried out by two American social psychologists, David Dunning and Justin Kruger, after whom the closely related Dunning-Kruger effect is named. This is another form of cognitive bias whereby people with low ability at a particular task overestimate their own ability while those with high ability overestimate the ability of others. David Dunning has explained it thus: "If you're incompetent, you can't know you're incompetent. . . . The skills you need to produce a right answer are exactly the skills you need to recognize what a right answer is."[20]

We see the Dunning-Kruger effect playing out every day on social media, particularly when it comes to conspiracy theories and irrational ideologies. Legitimate experts—whether scientists, economists, historians, lawyers, or serious

20 David Dunning, *Self-Insight: Roadblocks and Detours on the Path to Knowing Thyself* (New York: Psychology Press, 2005), 22.

journalists—tend to be far more ready to admit what they don't know than those with no particular training or knowledge of a subject beyond the superficial. This is partly why debate on social media has become so polarized and unproductive. Those most qualified to comment on a subject are also the most likely to be cautious and considered in the presentations of their arguments, because they know where there is less reliable evidence to bring to bear on the issue being discussed and where there are weaknesses in their understanding. (Often, they may simply choose not to engage in debates with people who determinedly value opinion over evidence.) And so they are more likely to remain silent, leaving a barren no-man's-land between the strident, uncompromising and, sadly, less well-informed warring factions on either extreme. Many studies have also found that, unlike those who are well-informed on a particular subject, those who are not so well-informed are also less willing, and less able, to acknowledge their weaknesses and hence the need to become better informed.

I should add a word of caution here that not everyone agrees that the Dunning-Kruger effect is even real;[21] it may just be an artefact of the data. But, the lesson to remember here is that we should not be so hasty to dismiss the views and questions of those who disagree with us because we think they are 'dumb'; we all should examine our own competences and biases before we critique those of others.

Of course, the paucity of considered and calm debate on social media is not just because those who are more knowledgeable on a subject are less likely to engage, for there are many issues on which the informed and the uninformed obviously do battle. But, because it is a part of human nature, confirmation bias is just as likely to afflict both sides—even if one side is objectively more 'right'. We are probably all guilty of this, however well-informed we *think* we are.

21 See, for example, Jonathan Jarry, "The Dunning-Kruger effect is probably not real", McGill University Office for Science and Society, December 17, 2020, https://www.mcgill.ca/oss/article/critical-thinking/dunning-kruger-effect-probably-not-real.

There is also a cultural element that feeds into issues of confirmation bias and illusory superiority. For example, studies[22] have shown that Americans who failed at a particular task were less inclined to persist on a follow-up task than those who had first succeeded, whereas the opposite was seen among Japanese subjects: those who first failed tried harder on the next task than those who had succeeded first time round.

Before we consider ways of tackling confirmation bias, let us examine whether the problem exists within science too. Why should scientists not be as likely as anyone else to succumb to confirmation bias? Surely everyone is susceptible to it. The problem is not, however, spread uniformly across all areas of science, and some disciplines suffer more than others. I hope I don't come across as biased myself when I say that in the

22 Steven J. Heine et al., "Divergent consequences of success and failure in Japan and North America: An investigation of self-improving motivations and malleable selves," *Journal of Personality and Social Psychology* 81. No. 4 (2001): 599–615, https://psycnet.apa.org/doiLanding?doi=10.1037%2F0022 -3514.81.4.599.

natural sciences, such as physics and chemistry and, to a lesser extent, biology, the problem is less prevalent than in the social sciences, where the complexity of the study of human behaviour means these sciences are more open to interpretation and subjectivity than the more exact natural sciences. This said, it certainly would be a good example of confirmation bias on my part, as a physicist, to assume that natural scientists are immune and can therefore let down their guard. In fact, social scientists are more familiar with the phenomenon due to the nature of their studies, and are therefore more conscious of and prepared to control for its insidious effects.

In the previous chapter, I discussed what is called the problem of inductive reasoning, whereby it is difficult to tell just how much evidence in support of a scientific theory is needed for the theory to be trusted. Often, a discovery is made, or evidence becomes available, which goes against established wisdom. If the evidence isn't overwhelmingly conclusive, scientists may ignore it, or cherry-pick those bits of it that suit their thinking. They may misinterpret, misunderstand, or even deliberately fabricate results, either to prop up an established theory that they favour or to promote their new one. Scientists are only human and suffer from the same weaknesses as everyone else, so there will always be the chance of bias at the individual level due to pride, jealousy, ambition or even just downright dishonesty.

Luckily, this is far rarer in science than you might think. Most such obstacles in the path of scientific progress are temporary, thanks to the built-in corrective mechanisms of the scientific method that acknowledge and act to reduce this bias, such as the requirement for reproducibility of results and the slow progress through consensus rather than diktat. Scientists also use a range of other techniques to rule out bias, such as randomised double-blind control trials (where not even the investigators know until the end which subjects received the intervention and which the placebo) and the peer review process of publication. Bad ideas in science never last very long— sooner or later the scientific method wins out and progress is made.

Sadly, everyday life is not so straightforward. An acquaintance of mine once told me he was convinced that aliens had visited Earth thousands of years ago and built the Pyramids of Giza using their advanced technology. He based this view not on the numerology associated with the Pyramids—the practice of looking for patterns and deeper meaning in the geometrical proportions of the pyramids' dimensions—but on the sheer precision with which the stone blocks fitted together. His thesis was that the blocks must have been cut by a laser, a technology that humans would not develop for a further four and half millennia. No amount of reasoning and persuasion on my part would shake him in his belief, which he had garnered from a number of YouTube documentaries he'd watched. No matter what evidence I offered to persuade him that archaeologists well understood how the stones had been cut, transported and raised into position, and why the Pyramids had been built; or how unlikely it was that aliens had visited us back then without leaving any reliable evidence or scientifically analysable traces—he continued to hold his view just as firmly. This 'belief perseverance' can be very powerful, especially if the believer can rationalise a way of viewing the evidence as supporting rather than opposing their beliefs.

How else can confirmation bias occur? You may have heard the phrase 'correlation does not imply causation', which means that if an association, or connection, is observed between two things this does not mean that one has to be the cause of the other. For example, it is a fact that, on average, cities with more churches tend to have more crime. That is, there is a strong positive correlation between the number of churches and number of crimes committed. Does this mean that churches are particularly good at turning people to crime, or maybe that a lawless city is more likely to need churches where criminals can confess their sins? Of course not. However, both effects are also correlated with a third parameter: the population of a city. All other things being equal, a city of five million people (in a predominantly Christian country) will contain more churches than a city of a hundred thousand people. It will also most likely record a higher number of crimes per year. The number of churches and number of crimes are correlated, but neither is the cause of the other. And yet many people will take such correlations at face value and erroneously deduce a causation without questioning the logic of their conclusion. Even when presented with the correct interpretation, such as a city's population in the above example, they still find it difficult to shake their initial inference (belief perseverance). This is sometimes referred to as 'the continued influence effect': believing a previously held view even after it has been shown to be wrong. It is most commonly seen in the various forms of misinformation spread by political campaigners, the tabloid media or bots on social media—once the germ of an idea has been planted, particularly if it fits with preconceived beliefs, it is very difficult to erase.

Since confirmation bias in its various forms is a part of human nature, you might argue that it is fruitless trying to tackle it by persuading others to think differently. So, what you can do instead is acknowledge that it most likely also exists in your own views. As the ancient Greek aphorism puts it, "Know thyself." Knowing this aspect of human nature means you can make some attempt to stand back and examine just why you hold those views and whether you are putting more weight on information you receive that confirms what you already think, while dismissing anything that contradicts it.

Ask yourself *why* you believe something to be true. Is it because you want it to be? Scientists are confident that anthropogenic climate change is happening, but contrary to what some people might think, the vast majority of scientists have no vested interest in believing that we are perilously changing our planet's climate. In fact, despite all the evidence, and unlike someone who denies anthropogenic climate change, we really hope we are wrong. After all, scientists also have children and grandchildren who will inherit the planet after they are gone.

So, when it comes to the hundreds of different topics that you might hold a strong opinion about, rather than leaping headlong into an argument with someone you disagree with, first take some time to examine your own motives for believing what you do and question the motives of the sources of your own information. Do you believe what you believe because it fits into your wider ideological, religious or political stance? Is it because others whose opinions you value also believe it? And, crucially, does that make it right? Finally, have you accessed sufficient relevant information and taken the time to establish that this information is reliable—and that you understand it? Once you have questioned your own beliefs, you can then begin to see things from a different perspective and establish whether your beliefs still make sense. It may be that you will still be convinced that you are right, and that is fine, as long as you have examined the evidence objectively. You may of course realise that you have more questions. And that's also okay. What is important is that you never stop this process of questioning what you believe, as doing so is exactly how you can clear away the fog of bias with the light of reason.

And then, what do you do if you find yourself persuaded that in fact, you have been wrong?

It's not always easy to admit this, even to your-self. There is another ancient Greek aphorism worth remembering at such times: "Surety brings ruin."

This brings us to the next chapter.

7

DON'T BE AFRAID TO CHANGE YOUR MIND

Recognising your biases is hard enough to do, but confronting them and acting to eliminate them is a different matter entirely. This often means that you must overcome the discomfort of admitting you may have been wrong about something and to be prepared to change your mind. The reason this is so difficult to do is due to what psychologists refer to as *cognitive dissonance*—a fascinating state of mind that arises when a person is faced with two conflicting viewpoints, typically a strongly held belief that butts up against newly acquired information that contradicts that belief. This leads to a feeling of mental discomfort that is most easily alleviated by dismissing the new information or downplaying its importance in order to hang on to what one believes to be true.

This is not the same phenomenon as cognitive bias, whereby a person is so sure they are right they do not even entertain any conflicting views in the first place.

We hear about cognitive dissonance more often these days as we sift through the ever-growing mountain of information we get buried in, because it is playing an increasingly important role in our decision-making process. It is not a new idea, nor particularly novel or controversial—cognitive dissonance has been well understood by psychologists for many years and is now very much a part of the popular zeitgeist alongside the idea of confirmation bias.

Can we tackle this problem too by thinking more scientifically? Let us first take a look at how science deals with it. I have already said that if scientists always held on to their preconceived ideas then they wouldn't make much progress. Of course, sometimes they have good reason for sticking to their guns, because the scientific theories they trust have been established through the slow and rigorous processes of the scientific method. Successful theories are those that have been tested, prodded and poked in a deliberate attempt to knock them off their pedestals. We gather data. We make observations. We carry out experiments. We develop models and theories that we hold up against rival ones to see which is the more accurate, reliable and predictive. If a theory survives, it is because it has been through this process of rigorous interrogation and we can be confident that the new scientific knowledge it gives us about the world can be trusted. And it is here where we find one of the most important features of the scientific method: all of these careful steps are built on acknowledging and quantifying uncertainty, because a good scientist will always retain some degree of doubt and rational scepticism. This does not necessarily mean that the scientist is sceptical of others' views, but rather that we as scientists should acknowledge that we may ourselves be wrong. The vital role of doubt and uncertainty in science means being open to new ideas and being prepared to change your mind when a deeper understanding is reached, or when better data or new evidence becomes available. Such an attitude avoids, or at least reduces, the problem of cognitive dissonance.

While doubt and uncertainty are important in science, so too is certainty; otherwise, we would likewise never make progress, and of course we do. The scientific method has many imperfections, and it is true that the *process* of scientific discovery is often messy and unpredictable, and full of blemishes, blunders and biases. But after the dust has settled on some aspect of our understanding of the world, we usually find that progress has been made not through doubt, but through well-founded conclusions based on carefully justified steps that gradually reduce our levels of uncertainty. I return to my favourite example: if I were to drop a ball from a height of five metres above the ground, then I am very, very sure—or rather, I have very, very little uncertainty—using a simple formula connecting distance, time and acceleration, that the ball will fall for one second before it hits the ground.

And yet uncertainty forms part of every theory, every observation, every measurement. A mathematical model will have built-in assumptions and approximations with a well-defined level of accuracy. Data points on a graph have error bars representing the level of confidence we have in their values—small error bars mean the values have been measured to a high degree of accuracy, whereas large error bars mean we are less confident. Measuring uncertainty and accepting it as an integral part of scientific investigation is ingrained into every science student.

The problem is that many people not trained in science see uncertainty as a weakness rather than a strength of the scientific method. They will say things like, "If scientists are not sure of their results and admit that there is a chance that they might be wrong, then why should we trust them at all?" Quite the opposite, in fact: uncertainty in science does not mean we don't know, but that we do know. We know just how likely our results are to be right or wrong because we can quantify our degree of confidence in them. To a scientist, 'uncertainty' means a 'lack of certainty'. It does not mean ignorance. Uncertainty leaves room for doubt, and this is liberating because it means we can critically and objectively assess what we believe. Uncertainty in our theories and models means that we know they are not absolute truths. Uncertainty in our data means our knowledge of the world is not complete. The alternative is far worse, for it is the blind conviction of the zealot.

Levels of confidence in scientific findings are also often misunderstood or misrepresented in the media. Sometimes this is the fault of the scientists themselves—for example, if they neglect to mention the level of uncertainty in their results in order for their discoveries to make the news and reach a wider audience. Similarly, if new products or technologies are promoted, then any uncertainties that could compromise commercial interests may be downplayed or ignored. Some journalists, often through no fault of their own but because they lack scientific training, can also disregard uncertainty by oversimplifying and cherry-picking words from a scientific paper or a press release. In doing so, they can misinterpret the often carefully chosen words of the authors—who in turn might themselves also be partially to blame for not anticipating such pitfalls.

How different this all is from the world of politics, wherein if you dither or show any hint of uncertainty in your arguments, this is interpreted as a sign of weakness. Voters can even regard certainty as a strength in their politicians since, as Berkeley professor of management Don A. Moore puts it, "Confident people inspire faith that they know what they're doing; after all, they sound so sure of themselves."[23] This attitude has crept into the wider public debate on political and social issues to such an extent that one is often not permitted to adopt the middle ground—all opinions must be firmly held at all times. This would not get you very far in science, since we must always be open to new evidence and to changing our minds in light of it. There is even a certain nobility in scientific culture in admitting one's mistakes.

23 Don A. Moore, "Donald Trump and the irresistibility of overconfidence", *Forbes*, February 17, 2017, https://www.forbes.com/sites/forbesleadershipforum/2017/02/17/donald-trump-and-the-irresistibility-of-overconfidence/?sh=784c50c87b8d

Making mistakes in science is how we improve our knowledge and increase our understanding of the world. Not admitting to our mistakes would mean never replacing current theories with better ones and never acknowledging revolutions in our understanding. Just like resisting certainty, admitting our mistakes is a strength of the scientific method, not a weakness. Just imagine for a moment how much better things would be if politicians had the honesty of scientists and admitted when they get things wrong. And lest you think I am singling out politicians, imagine how much healthier all debates and arguments would be if we were willing to concede when shown to be wrong. Reaching the truth of a matter should always take priority over point-scoring and winning arguments, however uncomfortable the cognitive dissonance may be.

Cognitive dissonance is not an unusual or atypical state of mind that needs 'curing'. Rather, it is a natural part of human nature, and we all experience it on some level. Life is full of conflicting thoughts and emotions, and it is the reason we argue with friends and loved ones, have doubts and regrets about decisions we make, do things we know we shouldn't, and so on. But just because it is human nature, that does not mean we shouldn't try to counter it. Cognitive dissonance is a sign that we are not thinking rationally, and we need to analyse our views and get back on the track of rationality if we want to make the right decisions in life. Cognitive dissonance makes us uncomfortable, and the easiest way to relieve this discomfort and remove the contradiction is to convince ourselves that we are making the right choices by disregarding or downplaying the evidence from the outside world that conflicts with our inner beliefs and emotions. What we should be doing instead is tackling our cognitive dissonance head on and analysing it logically. It may be less comfortable, but it will be more beneficial to us in the long run.

Now more than ever, we need to explore ways of dealing with cognitive dissonance, as it is far more serious in our modern culture and times than it has ever been. The spread of misinformation and the growing support for conspiracy theories come at a time when the world is facing huge challenges. For example, many people feel genuine cognitive dissonance when faced with a conflict between choosing to act on public health advice during a pandemic, when doing so restricts their freedoms, or choosing instead to follow their natural human urges to deny the evidence or downplay its importance, because of a desire to live in a less restricted way. Some may also feel very uncomfortable when the scientific community is advising one course of action and the government is advising another. These situations are incredibly challenging, but this is exactly when we need to take some time to analyse what we believe and why we believe it, as this will form the basis of our decisions—decisions guided by reason while at all times remaining open to change in the light of reliable new evidence.

Accepting that we might sometimes be wrong is how we can develop a deeper understanding of our world and our place in it—and it can be hugely rewarding, if we can manage it. As Oscar Wilde pointedly put it, "Consistency is the last refuge of the unimaginative." Breaking free of our desire for consistency and certainty is not always easy to do—and that goes for anyone—so it's helpful to break things down. Shake off your sense of surety. It may be uncomfortable at first, but you will adjust and actually find yourself *more* uncomfortable with those who profess certainty at all times. Listen to the views and arguments of the 'other side' with patience. Ask questions. Take time to find and understand evidence from reliable sources. Be wary of certainty, but let those who are open about (and better yet, can quantify) their own uncertainties earn your confidence. "Doubt is not a pleasant condition, but certainty is absurd," Voltaire once said. And remember: if you are wrong, be brave, be noble, and admit it—and value others who have the courage and integrity to do the same.

8

STAND UP FOR REALITY

The repercussions of the US presidential elections in 2020 will surely go down in history as a time when disinformation, driven largely by social media, came of age. For weeks after the November presidential elections in the US, many Americans who voted for Donald Trump refused to accept the result, which saw the Democrat candidate, Joe Biden, win decisively.[24] The accusations of fraud and cheating, mainly pushed out by President Trump himself over social media, were genuinely believed to be incontestable facts by millions of voters despite the lack of any credible evidence.

24 Based on all the evidence and information available to me, of course.

They were instead built on hearsay, rumour and outlandish conspiracy theories.

While this was going on, many more millions of people around the world subscribed to wild theories about the coronavirus pandemic: that the SARS-Cov2 virus was artificially produced in a lab in China/America[25] to control the world population, that it was spread by 5G networks and activated by face masks, or that powerful people such as billionaire Bill Gates were somehow involved in an international conspiracy to control our minds. And despite the hundreds of millions who were infected with the virus and the millions who died, there were still many who believed that the entire pandemic was fake news.

This phenomenon has been likened to a new form of solipsism, whereby many people inhabit their own parallel reality, built up from fake narratives and misinformation, superimposed onto the real one. But this is not quantum physics where all possible outcomes can be realised in

25 Depending on where those who subscribed to such conspiracy theories lived.

the multiverse. Our everyday reality is not like the world of subatomic particles. For us there is only room for one version of what is real.

Is this trend of people buying into preposterously false narratives disturbing? Of course. But is it surprising? No, not really—conspiracy theories are hardly a new phenomenon. However, the speed at which they can now spread, particularly over social media, is both remarkable and terrifying in equal measure.

Scientists pride themselves on believing that they are seekers of an objective truth about the world, but this is not always as straightforward as you might think, since obstacles such as confirmation bias and cognitive dissonance can afflict individual scientists the same as everyone else. But when it comes to trying to uncover the truth about some event or story in everyday life, then things can be even more complicated. For instance, a news report can be factually accurate and yet at the same time still be biased and subjective. In fact, different news networks, newspapers or websites can all report on the same event correctly and still have wildly deviating interpretations, with each stressing or exaggerating some aspects and downplaying others. They may not be deliberately trying to mislead or lie, but will simply be seeing the event or reporting the story through a lens defined by their ideological or political stance. Again, there is nothing new here, and if we are being diligent, then we will get our news from multiple sources in order to form a balanced view—although in reality few of us do this. However, when it comes to the spreading of pernicious fake news and deliberately misleading 'disinformation', as opposed to just plain 'misinformation' or biased reporting, then that is something we must try to combat.

False information—intentionally or inadvertently conveyed or spread—is not a consequence of the new digital technologies we have today, but it has certainly been amplified by them in recent years. So, what can be done about it? I discussed in the previous chapter how we can all question what we hear and read, by examining our biases and demanding hard evidence. But it is unlikely that any of this advice will truly change the mind of a conspiracy theorist. So, in the end it may just be that as a society we have to find ways to combat disinformation collectively, and that the laws and legislation we need to put in place must be tightened in order to stop lies and misinformation from leaking out and polluting our thoughts and opinions.

This problem is sadly becoming more acute by the day as the technologies used to spread information get ever more sophisticated. We are already at the stage where it is almost impossible to distinguish between fake and real images, video footage or audio, and it is becoming increasingly easy with widely available technology to create and disseminate fake facts. At the same time, the technology used to distinguish between what is fake and what is real is still too easily fooled. We will therefore have to find ways and develop strategies to deal with misinformation and fake narratives as quickly as possible. These will require a combination of technological solutions and societal and legal changes.

When we hear about the use of AI algorithms and machine learning today it tends to be in the somewhat negative context of filtering information to make us more easily targeted by advertisers. Far more insidious, however, is the fact that this technology is also being used to spread misinformation by making the fake almost indistinguishable from the real thing. But AI can also be used for good. It can be used to do the checking, assessing and filtering for us. We will soon develop advanced algorithms to identify, block or remove online content that is found to be false or misleading.

We are therefore now witnessing technological advances pulling in opposite directions. While it is increasingly easy to create convincing fake information, it is also possible to use the same technology to verify that information. Which of these two competing forces (good versus evil) wins in the end is down to us and how we respond.

Pessimists will naturally ask whose truth we will end up with. Some may even argue that personal freedom should be valued above truth. They will say that increased censorship and mass surveillance will tend to create official 'truths' that society has to sign up to. Or they may be concerned that the technologies used to filter out the fake information are put in place by powerful bodies, such as Facebook and Twitter, that are not themselves entirely objective and may have their own vested interests and political ideologies.

It is surely encouraging to see many of the big social media platforms now starting to deploy increasingly sophisticated algorithms to deal with content online deemed by wider society to be morally undesirable, such as incitement to violence, dangerous ideologies, racism, misogyny and homophobia, as well as information that can be proven evidentially to be fake. However, 'outsourcing' this responsibility to private and hugely powerful entities, which mainly exist, after all, to make a profit, may not be entirely desirable in the long term. And if we must use them then we will have to find clear ways of holding such bodies accountable for the steps they are taking on our behalf.

It is even contended that *any* system with the ability to 'judge' information as true or false is inherently biased. However, while it is true that the design of these systems is being done by humans who themselves have values and biases, this is taking the argument too far, and I personally don't buy it. As AIs become more sophisticated, they can certainly help us to weed out falsehoods and identify evidence-based facts, but they can also highlight where there is uncertainty, subjectivity and nuance. In a well-known comedy sketch on British TV, someone playing the role of customer service staff relying on decisions generated by a computer always responds to the most reasonable of customer requests with the phrase, "The computer says no". Technology has moved beyond this now. Recent advances mean that AIs will soon be able to embed moral and ethical thinking into their algorithms, enabling them to protect rights such as free speech while at the same time screening and blocking false narratives and misinformation. We need to control for bias, and so precisely what morals or ethics are encoded into these algorithms is something we must discuss openly and collectively as a society. What about religious versus secular beliefs? What about cultural norms? What some in society see as an acceptable, even necessary, moral standard, others regard as taboo.

There will always be those who mistrust any attempts to filter the truth from the lies. In a sense, this is unavoidable. It is not admitting defeat, but simply facing up to reality. We cannot hope to persuade and convince everyone—but we do have a responsibility as a society to try to ensure that no one intent on spreading lies and misinformation for their own nefarious ends ever gets to be in a position of influence, since this can have far-reaching consequences and potentially alter the future course of humanity. Throughout history, there have been despotic rulers, unpleasant political leaders and false prophets who through force, coercion and lies have convinced millions to follow them. Such people will always be with us. But what we can try to do is stop them from using science and technology as a weapon to advance their agenda.

What then is the lesson here? I have tried to end each chapter on a positive note, but I have painted a rather gloomy picture in this one. So, what hope is there in the future for truth to prevail over lies? It is widely acknowledged that things are likely to get worse before they get better, but we are developing the tools to solve this problem. For example, we can again take a leaf out of the scientific method. When information claims to come with supporting evidence, there will be the need to assess the quality of that evidence, for example by attaching a 'confidence level' to it. Thus, along with stating any claim, we should also try to include the uncertainty associated with it. Every scientist knows how to put error bars on data points; we need to do a similar thing—in spirit if not literally—when confronted with new information. To do this we will need to develop AI technologies that can provide us with a 'trust index' that shows how the veracity of information is linked to how trustworthy or credible the source of that information is perceived to be. If a source—whether a news outlet, website or even an individual social media 'influencer'—is marked as disseminating fake news, then that source will sit lower down on the trust index.

We are also making advances in what are known as *semantic technologies*, the aim of which is to help AIs interpret and truly understand the data by encoding meaning separately from the application code. Semantic technologies differ radically from the way machines traditionally interpret data, whereby meaning and relationships are hard-wired into the coding by human programmers. Like machine learning, semantic technologies take us along the path towards artificial 'intelligence' in the true sense of the word.

However, just as the problem of fake news and misinformation is not the fault of technology alone, neither is the solution going to be found through technological advances alone. This is really a societal problem that technology has amplified, and as such it requires societal solutions too. The statistician David Spiegelholter says that the biggest predictor of people's resilience to misinformation is numeracy. What he means is that it helps if we have some level of understanding and appreciation of data and statistics—what is called information literacy. One problem is that the media and politicians are not trained to communicate data and results clearly and accurately, and so they also need to be able to recognize when information is needed and how to get hold of it, evaluate it and use it effectively. So rather than rely entirely on clever technology to tell us what to believe and what not to believe, we all need to learn how to exercise better critical thinking skills ourselves. To do this we must address these fundamental skills in our education system. Therefore, alongside the exciting, shiny technology, we also need to learn better civics, better critical thinking skills and better information literacy.

As a society we must all learn to apply the methods of science: to develop mechanisms for coping with complexity, to assess uncertainty—to keep an open mind about information we only have partial knowledge about. While it is sad but undeniably true that a significant fraction of the population has neither the skills nor the ability to cope with growing complexity, it is also the case that ignorance often leads to frustration, disillusionment and helplessness, all of which provide a fertile breeding ground for the growth and spread of misinformation and fake narratives. These problems have always been with us and always will be. It is human nature to gossip, to fabricate, to exaggerate, while those in power will always use propaganda or distortion of the truth for political or financial purposes. But we cannot deny that these problems have become more acute with advances in technology.

I have always been an optimist and tend to have faith in humanity's good side. Humankind has always found ways to overcome its problems through innovation and ingenuity—by and large making the world a better rather than a worse place.[26] So, I am confident that we will find solutions, whether technological or through better education. However, if we are to succeed, we need motivation and fortitude. We must stand up for reality, for truth. We must learn good judgement, develop our analytical skills, help our loved ones do the same, and expect the same of

26 A good book on this is Steven Pinker's *The Better Angels of Our Nature: Why Violence Has Declined* (New York: Viking, 2011).

our leaders. We must all . . . well . . . think more scientifically. This is how we can better understand and withstand the challenges that the real world throws at us and make better decisions in our lives. This is how we can also stand up for the reality that we want for ourselves and others—a world in which we are no longer prisoners, chasing shadows in the dark, but are freer and more enlightened.

CONCLUSION

In this book, I have considered how we can live a more rational life. But what is the true value to humanity of thinking scientifically? There are, in my view, four answers.

Firstly, in developing the scientific method, humankind created a reliable way of learning how the world works, a way that takes into account our human foibles and builds in correctives. I see this as the intrinsic value that the mode of scientific thinking possesses. Investigating the world through a scientific approach has revealed profound truths that will never be overthrown. Consider one of the biggest ideas in my own field of physics. Einstein's theory of gravity replaced Newton's to give us a more accurate and more fundamental account of the structure of the cosmos. And while we cannot rule out the possibility that Einsteinian relativity may itself be replaced one day by an even deeper theory, this will never alter the fact that Earth orbits the Sun, not the other way round, and that the Sun is one of hundreds of billions of stars in our Milky Way Galaxy, which is itself one of many billions of galaxies in the known universe. And is it not inspiring that we can share not only the things we have learned about our world—across the globe and across time—but also a *way* of thinking and learning? Because this means that, even if the record of all knowledge itself is lost, we can still use the scientific method to rebuild it over time.

Maybe this means of gaining knowledge and understanding that science has given us is not as inspiring to you as it is to me, but there can be no denying the second reason we should value it. We trust in science *because it works* and because we recognise where we would be without it. When people ask me why I am so convinced that a theory as crazy and non-intuitive as quantum mechanics is correct, I ask: Do you like your smartphone? Are you not astonished by what it can do? Well, you have quantum mechanics to thank for its existence. Your smartphone, and every other modern electronic device you are familiar with, is packed with technology that is only possible because of our understanding of the behaviour of matter at the tiniest scales, which we have gained through the development and application of the theory of quantum mechanics. So, the theory may seem utterly perplexing and bizarre to us, but it works.

Too many people have lost sight of the way science and technology are interwoven. This is partly because scientists themselves have tended to separate the two. Science is the *creation* of knowledge, we have argued, whereas technology is the *application* of that knowledge. But this sharp distinction doesn't always make sense; after all, most scientific work is about more than learning something we didn't previously know. Would we not call the mixing of chemicals, whether in a school lab or an industrial lab, doing 'science'? And would we not refer to work that applies existing knowledge to designing a more efficient laser or to developing a better vaccine as 'science'? In all these examples, we are not gaining new knowledge about the world, and so such a narrow definition of what it means to be doing science is wrong. Applied science is still science.

And yet we do claim that science is value-neutral—that it is neither good nor bad—and it is how we put it to use that can sometimes be the problem. Einstein's equation, $E = mc^2$, is simply a fact about our universe that links mass and energy via the speed of light. But using it to build an atom bomb is a different matter entirely. Would it have been better if Einstein had never discovered relativity theory? Would that mean that atom bombs would never have been dropped on Hiroshima and Nagasaki? Well, leaving aside the argument that had Einstein not discovered relativity someone else would have done so soon after anyway, can it be better to 'un-know' something about the world? Of course not. Yes, this is an example of when scientific knowledge has given humankind the potential to do evil. But that is *not* the same as saying that the scientific knowledge itself is evil, or that unknowing it would have made for a better world.

Without science we would not have the means to feed a growing global population, to live longer and happier lives, to light and heat our homes, to communicate with each other, to

travel the world and beyond, to build great ci-
vilisations and democracies, to understand our
bodies and to develop the drugs and vaccines
that protect us from disease, to relieve millions
from the burden of hard manual labour and free
more of us to enjoy art, literature, music and
sport. Without science, there would be no modern world—and, we might argue, no future for
our species. So, we should not forget that science is more than a pursuit of knowledge. It is
a means by which we can both survive and live
more contented lives.

The third value of thinking scientifically has
been the theme of this book. The way we do
science—all the traits and practices of science,
such as being curious about the world, thinking
rationally and logically, debating, discussing
and comparing ideas, valuing uncertainty and
questioning what we know or think we know,
acknowledging our biases, demanding reliable
evidence, learning what and whom to trust—all
these can benefit us in our everyday lives. The
more we understand, the more enlightened we
are and the better positioned we are to make

rational decisions that will help us and those we care about.

And yet . . . there is one last value of thinking scientifically that I wish to end with. I would argue that, for all the breadth and complexity of our scientific knowledge to date—and it is very far from complete, nor will it ever be—and for all the remarkable technological, social and medical advances science has given us, and for all the messy, rich, complicated splendour of the scientific method we have used to gain this knowledge, the true beauty of science is that it enriches us. As Carl Sagan says, the combined "sense of elation and humility" it gives us, is "surely spiritual".

We are a species that has had remarkable evolutionary success over time. Our collective knowledge gives us enormous power and potential. Yet, we are fragile. We are fractious. The scientific knowledge we have accumulated, and the technologies we continue to develop through science, have not been shared widely or equally. Yet, the scientific approach—this remarkable way of seeing, thinking, knowing, and living—is one of humankind's great riches and the birthright of everyone. And, most wonderfully, it only grows in quality and value the more, and the more widely, it is shared.

Science is so much more than hard facts and lessons in critical thinking, just as a rainbow is so much more than just a pretty arc of colour. Science gives us a way to see the world beyond our limited senses, beyond our prejudices and biases, beyond our fears and insecurities, beyond our ignorance and weaknesses. Science helps us to see through a lens of deeper understanding and be part of a world of light and colour, of beauty and truth.

The next time you see a rainbow, you will know something not everyone around you does. Would you keep that a secret from the person standing next to you? Do you think telling them what you now know about it would ruin the magic? Or would it be a joy to share that knowledge?

You won't find a pot of gold at the end of a rainbow—remember, a rainbow doesn't really have an end. But you can find hidden riches within yourself—in the enlightened way of thinking and seeing the world that you can now embody and make use of in your daily life and share with those you know and love. That is the wonder. That is the joy of science.

GLOSSARY

Allegory of the cave

An allegory on the importance of education over ignorance presented by the Greek philosopher Plato around 375 BC in his Socratic dialogue *The Republic.* In it he describes how a prisoner, freed from being chained in a cave, comes to see a higher level of reality outside.

Belief perseverance

The tendency to stubbornly cling to one's initial belief even after receiving new information that firmly contradicts the basis of that belief.

Cognitive dissonance

The feeling of mental discomfort when a person is faced with two contradictory ideas or beliefs—typically, a preexisting, strongly held one which conflicts with newly acquired information. The discomfort is most easily alleviated through belief perseverance (see above): by dismissing the new information or downplaying its importance in order to hang on to what one already believes to be true.

Confirmation bias
The tendency to expose oneself only to those opinions and beliefs that confirm what one already thinks and to accept only the evidence that supports this.

Conspiracy theory
In general, a conspiracy theory is an explanation for a phenomenon or event that rejects the standard accepted explanation in favour of one claimed to be the 'truth' that has been covered up or suppressed for covert or sinister reasons by organisations, governments or powerful vested interest groups. Rejected explanations include those that have the backing of mainstream scientific evidence.

Conspiracy theories resist falsification. Instead, any evidence against the conspiracy, or absence of evidence supporting it, is often reinterpreted as evidence of its truth. This distinguishes conspiracy theories from scientific ones in that they become more a matter of faith than reason, despite their advocates firmly believing that they have ample supporting evidence and are thinking rationally.

Cultural relativism

Culture is the set of beliefs, behaviours or characteristics shared by a group of people or a society as a whole, which are based on traditions, customs and values. Relativism is the view that whether something is true or false, right or wrong, acceptable or unacceptable, is relative—that there is no reference frame or vantage point from which to establish an objective and absolute answer that all can agree on.

At its most basic, and positive, cultural relativism can be viewed as a general tolerance and respect for difference whereby culture and customs are not judged to our own standards of what is right or wrong, strange or normal. Instead, we should try to understand cultural practices of other groups within their own cultural context.

However, it is when relativism clashes with realism that problems can arise. This was discussed in the eighteenth century by Immanuel Kant in his *Critiques,* in which he argued that our experience of the world is mediated through the knowledge and ideas we hold. For example, if cultural relativism's contention is that there is no such thing as universal, objective moral truth, we should be careful not to allow this notion to contaminate our rational thinking about objective reality and scientific truth.

Disinformation

A type of misinformation that is spread deliberately to deceive or mislead.

Dunning-Kruger effect

A type of cognitive bias described by social psychologists David Dunning and Justin Kruger in which people with limited knowledge or competence believe that they are actually smarter and more capable than they really are. This combination of low cognitive ability and poor self-awareness means that they are unable to recognise their own shortcomings. Conversely, those with high competence tend to underestimate their own ability because they fail to recognise the incompetence of others.

However, the Dunning-Kruger effect has been challenged by studies that suggest it is no more than an artefact of the data.

Falsifiability

A scientific theory is falsifiable (or refutable) if it can be contradicted by a logically possible observation. This concept was introduced by the philosopher of science Karl Popper as the Falsification Principle, a way of determining whether a theory or hypothesis is scientific or not. To qualify, it must be testable and potentially disprovable.

Illusory superiority

A condition of cognitive bias wherein a person will overestimate their own competence and ability in relation to the same qualities in other people. It is related to the Dunning-Kruger effect.

Implicatory denial

One of the three forms of denial described by the late psychoanalytic sociologist Stanley Cohen. Here, it is not the facts themselves that are denied, but rather their implications and consequences. The oft-quoted example is climate change, whereby it is acknowledged that it is indeed taking place, and even that it is due to humanity's actions, but what is denied is its moral, social, economic or political implications, thereby removing any responsibility or need to act.

Interpretive denial

Here, the facts themselves are not denied, but rather interpreted in ways that diminish their importance or distort their meaning. For example, it is not denied that the climate is changing, but rather that the rise in temperature is due to natural solar cycles and the increase in greenhouse gases is a consequence, not a cause, of this.

Literal denial

The outright rejection that something has happened or is happening, usually despite solid evidence to the contrary. Such denial can be deliberate (perhaps for ideological reasons) or through disinformation and ignorance. The best-known example is Holocaust denial.

Misinformation

False or misleading information that is communicated whether or not there is any deliberate intention to deceive. Examples include gossip or hearsay based on ill-informed opinion, anecdotal evidence not supported by good data, poor journalism, political propaganda or even, in some cases, deliberate lies created to serve some ulterior purpose (disinformation).

Moral truth

We usually say that a statement is 'true' when it corresponds with reality, or with the way the world 'really is'. In philosophy, this is known as the correspondence theory of truth—that truth corresponds to objective facts. With moral truths, the water is muddier. The existence of absolute moral truths depends on whether one believes that there are universal ethical standards applying across all contexts, cultures, times and people—for example, that murder is bad. Such moral truths are then grounded in ethical law or in religious texts or are adhered to uncompromisingly due to strongly held beliefs or upbringing. In contrast, relative moral truths (moral relativism) are subjective and context dependent (for example, polygamy is frowned upon is many societies but tolerated or deemed acceptable in others). However, even such definitions are not particularly helpful, since what one person might regard as an absolute moral truth might be seen by someone else as relative.

Ockham's (or Occam's) razor

Sometimes referred to as the principle of parsimony, this is the idea that the simplest explanations are usually the best ones, or that one should not make more assumptions than absolutely necessary.

Objective reality

The idea that the external physical world exists independently of the mind. While what we perceive may never quite be 'ultimate' reality, there is still a real world 'out there' whether or not we can ultimately know it fully. Its existence has been the subject of serious debate ever since it was called into question regarding the meaning of quantum mechanics in the 1920s. That debate is ongoing in the philosophy of physics.

Post-truth

The calling into question of facts and expert opinions, which are relegated to being of secondary importance to appeals to emotion through the repetition of unproven assertions. It has been argued that an early form of post-truth arose in the seventeenth century with the invention of the printing press and the so-called pamphlet wars. A subset of the idea is the modern culture of post-truth politics (also known as post-factual politics), which has arisen in the late twentieth and early twenty-first centuries in many countries, largely accelerated by the internet and social media, in which populist political debate is framed by appeals to emotion rather than fact.

Precautionary principle
The general philosophical and legal approach to policies or innovations that have the potential for causing harm by erring on the side of caution, particularly if compelling scientific evidence on the matter is still lacking.

Problem of induction
Induction is a type of scientific inference whereby a conclusion is reached based on the accumulation of observational evidence. Its weakness (the problem of induction) is that we cannot know how much evidence is sufficient and of what quality it must be in order to reach a firm conclusion.

Randomised control trial
A scientific method used to study causal relationships in order to minimise bias. Typically, a statistically significant number of similar people are randomly assigned to two groups, for example to test a new medical treatment, drug or intervention. One group (the experimental

group) receives the intervention being tested, while the other (the control group) receives an alternative intervention, usually a dummy (placebo) or even no intervention at all. It is also usually 'double blind' because not until a study is completed do the investigators themselves know who was in which group. The difference in response between the two groups is analysed statistically to test the efficacy of the intervention.

Reference frame independence

The scientific concept, used mainly in physics, whereby some quantity or phenomenon has a fixed value or property when viewed from different frames of reference or perspectives. The most famous example is the value of the speed of light in a vacuum, which, unlike the speed of material bodies, does not depend on the speed of the observer measuring it. More generally, the idea of reference frame independence can be applied to an external objective reality, which scientists try to understand independently of their subjective experience.

Reproducibility

In the scientific method, reproducibility refers to the degree of agreement between the results of experiments conducted by different individuals, at different locations, with different instruments. It is thus a measure of scientists' ability to replicate the findings of others, which, if successful, builds trust in those findings.

Reproducibility is distinct from repeatability, which measures the variation in results under the same conditions; that is, taken by the same instrument, in the same location, following the same procedure, by the same person and repeated over a short period of time.

Scientific truth

Scientists and philosophers have long argued about whether scientific truth even exists. Some think of it as a Platonic ideal that can never be reached, and may not even exist. Others insist that the true nature of reality, whether we can ever fully comprehend it or not, does exist, and that it is the job of science to try to approximate this so-called 'scientific truth' as closely as it can, through explanations, theories and observations. Note that what is meant by scientific truth is not the same thing as moral truth or, say, religious truth.

Scientific uncertainty

A term referring to the range of possible values within which a measurement lies. It provides us with a level of confidence in an observation or measurement, or in the accuracy of a theory. Further measurements that are more careful or additional refinements to the theory can reduce this uncertainty. A related term is the 'error' in a measurement, which does not mean that the measurement is wrong, but rather refers to the 'margin of error'. All scientists are trained to add 'error bars' to their data points to quantify the uncertainty.

Social construct

Something that is built up as a result of human interactions and shared experiences rather than existing as an independent objective reality.

Thus, while the scientific method itself is a social construct, the scientific knowledge about the world that it helps us accumulate is not.

The scientific method

A way of acquiring knowledge about the world that has been the hallmark of how science is conducted ever since the seventeenth century and the birth of modern science, most notably due to the work of the likes of Francis Bacon and René Descartes. However, its roots go back to the eleventh century and the Arab scholar Ibn al-Haytham. It involves the development of a hypothesis, testing it against careful observation and measurement, and the application of rigorous scepticism about what is claimed or observed. The practice of the scientific method requires honesty, the elimination of bias, repeatability, falsifiability and the acknowledgement of uncertainty and mistakes. It is the most reliable way we have of learning about the world because it has many built-in corrective mechanisms that compensate for subjectivity and for human errors and frailties.

Value neutrality

This is the state scientists attempt to achieve regarding their research in which they are objective, impartial and not influenced by their personal values or beliefs. Whether science can ever be truly value-neutral is a topic of continuing debate; while individual scientists cannot be entirely value-neutral, however hard they may try, there are certainly facts about the external physical world (see Scientific Truth and Objective Reality) that *are* value-neutral, such as the structure of DNA or the size of the Sun relative to the Earth.

BIBLIOGRAPHY

Aaronovitch, David. *Voodoo Histories: The Role of the Conspiracy Theory in Shaping Modern History.* New York: Riverhead Books, 2009.

Allington, Daniel, Bobby Duffy, Simon Wessely, Nayana Dhavan, and James Rubin. "Health-protective behaviour, social media usage and conspiracy belief during the COVID-19 public health emergency." *Psychological Medicine* 1–7 (2020). https://doi.org/10.1017/S003329172000224X.

Anderson, Craig A. "Abstract and concrete data in the perseverance of social theories: When weak data lead to unshakeable beliefs." *Journal of Experimental Social Psychology* 19, no. 2 (1983): 93–108. https://doi.org/10.1016/0022-1031(83)90031-8.

Bail, Christopher A., Lisa P. Argyle, Taylor W. Brown, John P. Bumpus, Haohan Chen, M. B. Fallin Hunzaker, Jaemin Lee, Marcus Mann, Friedolin Merhout and Alexander Volfovsky. "Exposure to opposing views on social media can increase political polarization." *PNAS* **115**, no. 37 (2018): 9216–21. https://doi.org/10.1073/pnas.1804840115.

Baumberg, Jeremy J. *The Secret Life of Science: How It Really Works and Why It Matters.* Princeton, NJ: Princeton University Press, 2018.

Baumeister, Roy F., and Kathleen D. Vohs, eds. *Encyclopedia of Social Psychology.* Thousand Oaks, CA: SAGE Publications, 2007.

Bergstrom, Carl T., and Jevin D. West. *Calling Bullshit: The Art of Scepticism in a Data-Driven World.* London: Penguin, 2021.

Boring, Edwin G. "Cognitive dissonance: Its use in science." *Science* **145**, no. 3633 (1964): 680–85. https://doi.org/10.1126/science.145.3633.680.

Boxell, Levi, Matthew Gentzkow, and Jesse M. Shapiro. "Cross-country trends in affective polarization." *NBER Working Paper* no. w26669 (2020). Available at SSRN: https://ssrn.com/abstract=3522318

———. "Greater Internet use is not associated with faster growth in political polarization among US demographic groups." *PNAS* **114**, no. 40 (2017): 10612–17. https://doi.org/10.1073/pnas.1706588114.

Broughton, Janet. *Descartes's Method of Doubt*. Princeton, NJ: Princeton University Press, 2002. htttp:www.jstor.org/stable/j.ctt7t43f.

Cohen, Morris R., and Ernest Nagel. *An Introduction to Logic and Scientific Method*. London: Routledge & Sons, 1934.

Cohen, Stanley. *States of Denial: Knowing About Atrocities and Suffering*. Cambridge, UK: Polity Press, 2000.

Cooper, Joel. *Cognitive Dissonance: 50 Years of a Classic Theory*. Thousand Oaks, CA: SAGE Publications, 2007.

d'Ancona, Matthew. *Post-Truth: The New War on Truth and How to Fight back*. London: Ebury Publishing, 2017.

Domingos, Pedro. "The role of Occam's razor in knowledge discovery." *Data Mining and Knowledge Discovery* 3 (1999): 409–25. https://doi.org/10.1023/A:1009868929893.

Donnelly, Jack, and Daniel J. Whelan. *International Human Rights*. 6th ed. New York: Routledge, 2020.

Douglas, Heather E. *Science, Policy, and the Value-Free Ideal.* Pittsburgh: University of Pittsburgh Press, 2009.

Dunbar, Robin. *The Trouble with Science.* Reprint ed. Cambridge, MA: Harvard University Press, 1996.

Dunning, David. *Self-Insight: Roadblocks and Detours on the Path to Knowing Thyself.* Essays in Social Psychology. New York: Psychology Press, 2005.

Festinger, Leon. "Cognitive dissonance." *Scientific American* **207**, no. 4 (1962): 93–106. http://www.jstor.org/stable/24936719.

———. *A Theory of Cognitive Dissonance.* Reprint ed. Redwood City, CA: Stanford University Press, 1962. First published 1957 by Row, Peterson & Co. (New York).

Goertzel, Ted. "Belief in conspiracy theories." *Political Psychology* **15**, no. 4 (1994) : 731–42. www.jstor.org/stable/3791630.

Goldacre, Ben. *I Think You'll Find It's a Bit More Complicated Than That.* London: 4th Estate, 2015.

Harris, Sam. *The Moral Landscape: How Science Can Determine Human Values.* London: Bantam Press, 2011.

Head, Megan L., Luke Holman, Rob Lanfear, Andrew T. Kahn, and Michael D. Jennions. "The extent and consequences of p-hacking in science." *PLoS Biology* **13**, no. 3 (2015): e1002106. https://doi.org/10.1371/journal .pbio.1002106.

Heine, Steven J., Shinobu Kitayama, Darrin R. Lehman, Toshitake Takata, Eugene Ide, Cecilia Leung, and Hisaya Matsumoto. "Divergent consequences of success and failure in Japan and North America: An investigation of self-improving motivations and malleable selves." *Journal of Personality and Social Psychology* **81**, no. 4 (2001): 599–615. https://doi .org/10.1037/0022-3514.81.4.599.

Higgins, Kathleen. "Post-truth: A guide for the perplexed." *Nature* **540** (2016): 9. https:// www.nature.com/news/polopoly_fs/1.21054! /menu/main/topColumns/topLeftColumn /pdf/540009a.pdf.

Isenberg, Daniel J. "Group polarization: A critical review and meta-analysis." *Journal of Personality and Social Psychology* **50**, no. 6 (1986): 1141–51. https://doi.org/10.1037/0022 -3514.50.6.1141.

Jarry, Jonathan. "The Dunning-Kruger effect Is probably not real." McGill University Office for Science and Society, December 17, 2020. https://www.mcgill.ca/oss/article/critical -thinking/dunning-kruger-effect-probably -not-real.

Kahneman, Daniel. *Thinking, Fast and Slow.* London: Allen Lane, 2011. Reprint: Penguin, 2012.

Klayman, Joshua. "Varieties of confirmation bias." *Psychology of Learning and Motivation* **32** (1995): 385–418. https://doi.org/10.1016 /S0079-7421(08)60315-1.

Klein, Ezra. *Why We're Polarized.* New York: Simon & Schuster, 2020.

Kruger, Justin, and David Dunning. "Unskilled and unaware of it: How difficulties in recognizing one's own incompetence lead to inflated self-assessments." *Journal of Personality and*

Social Psychology 77, no. 6 (1999): 1121–34. https://doi.org/10.1037/0022-3514.77.6.1121.

Kuhn, Thomas S. *The Structure of Scientific Revolutions*. 50th anniversary ed. Chicago: University of Chicago Press, 2012.

Lewens, Tim. *The Meaning of Science: An Introduction to the Philosophy of Science*. London: Penguin Press, 2015.

Ling, Rich. "Confirmation bias in the era of mobile news consumption: The social and psychological dimensions." *Digital Journalism* 8, no. 5 (2020): 596–604. https://doi.org/10.1080/21670811.2020.1766987.

Lipton, Peter. "Does the truth matter in science?" *Arts and Humanities in Higher Education* 4, no. 2 (2005):173–83. https://doi.org/10.1177/1474022205051965; Royal Society 2004; Medawar Lecture, "The truth about science." *Philosophical Transactions of the Royal Society B* **360**, no. 1458 (2005): 1259–69. https://royalsocietypublishing.org/doi/abs/10.1098/rstb.2005.1660.

———. "Inference to the best explanation." In *A Companion to the Philosophy of Science*, edited by W. H. Newton-Smith, 184–93. Malden, MA: Blackwell, 2000.

MacCoun, Robert, and Saul Perlmutter. "Blind analysis: Hide results to seek the truth." *Nature* **526** (2015): 187–89. https://doi.org/10.1038/526187a.

McGrath, April. "Dealing with dissonance: A review of cognitive dissonance reduction." *Social and Personality Psychology Compass* **11**, no. 12 (2017): 1–17. https://doi.org/10.1111/spc3.12362.

McIntyre, Lee. *Post-Truth*. Cambridge, MA: The MIT Press, 2018.

Nickerson, Raymond S. "Confirmation bias: A ubiquitous phenomenon in many guises." *Review of General Psychology*. **2**, no. 2 (1998): 175–220. https://doi.org/10.1037/1089-2680.2.2.175.

Norgaard, Kari Marie. *Living in Denial: Climate Change, Emotions, and Everyday Life*. Cambridge, MA: The MIT Press, 2011. *JSTOR*: http://www.jstor.org/stable/j.ctt5hhfvf.

Oreskes, Naomi. *Why Trust Science?* Princeton, NJ: Princeton University Press, 2019.

Pinker, Steven. *The Better Angels of Our Nature: Why Violence Has Declined*. New York: Viking Books, 2011.

Popper, Karl R. *The Logic of Scientific Discovery*. London: Hutchinson & Co., 1959; London and New York: Routledge, 1992. Original title: *Logik der Forschung: Zur Erkenntnistheorie der modernen Naturwissenschaft*. Vienna: Julius Springer, 1935.

Radnitz, Scott, and Patrick Underwood. "Is belief in conspiracy theories pathological? A survey experiment on the cognitive roots of extreme suspicion." *British Journal of Political Science* 47, no. 1 (2017): 113–29. https://doi.org/10.1017/S0007123414000556.

Ritchie, Stuart. *Science Fictions: Exposing Fraud, Bias, Negligence and Hype in Science*. London: The Bodley Head, 2020.

Sagan, Carl. *The Demon-Haunted World: Science as a Candle in the Dark*. New York: Random House, 1995.Reprint, New York: Paw Prints, 2008.

Scheufele, Dietram A., and Nicole M. Krause. "Science audiences, misinformation, and fake news." *PNAS* **116**, no. 16 (2019): 7662–69. https://doi.org/10.1073/pnas.1805871115.

Schmidt, Paul F. "Some criticisms of cultural relativism." The Journal of Philosophy 52, no. 25 (1955): 780–91. https://www.jstor.org/stable/2022285.

Tressoldi Patrizio E. "Extraordinary claims require extraordinary evidence: The case of non-local perception, a classical and Bayesian review of evidences." *Frontiers in Psychology* **2** (2011): 117. https://www.frontiersin.org/articles/10.3389/fpsyg.2011.00117/full.

Vickers, John. "The problem of induction." The Stanford Encyclopaedia of Philosophy, Spring 2018. https://plato.stanford.edu/entries/induction-problem/.

Zagury-Orly, Ivry, and Richard M. Schwartzstein. "Covid-19—A reminder to reason." *New England Journal of Medicine* **383** (2020): e12. https://doi.org/10.1056/NEJMp2009405.

FURTHER READING

Jim Al-Khalili, *The World According to Physics* (Princeton University Press, 2020)

Chris Bail, *Breaking the Social Media Prism: How to Make Our Platforms Less Polarizing* (Princeton University Press, 2021)

Jeremy J. Baumberg, *The Secret Life of Science: How It Really Works and Why It Matters* (Princeton University Press, 2018)

Carl Bergstrom and Jevin West, *Calling Bullshit: The Art of Scepticism in a Data-Driven World (Penguin, 2021)*

Richard Dawkins, *Unweaving the Rainbow: Science, Delusion and the Appetite for Wonder* (Allen Lane, 1998)

Robin Dunbar, *The Trouble with Science* (Harvard University Press, 1996)

Abraham Flexner and Robert Dijkgraaf, *The Usefulness of Useless Knowledge* (Princeton University Press, 2017)

Ben Goldacre, *I Think You'll Find It's a Bit More Complicated Than That* (4th Estate, 2015)

Sam Harris, *The Moral Landscape: How Science Can Determine Human Values* (Bantam Press, 2011)

Robin Ince, *The Importance of Being Interested: Adventures in Scientific Curiosity* (Atlantic Books, 2021)

Daniel Kahneman, *Thinking, Fast and Slow* (Penguin, 2012)

Tim Lewens, *The Meaning of Science: An Introduction to the Philosophy of Science* (Penguin Press, 2015)

Naomi Oreskes, *Why Trust Science?* (Princeton University Press, 2019)

Steven Pinker, *Enlightenment Now: The Case for Reason, Science, Humanism, and Progress* (Penguin, 2018)

Steven Pinker, *Rationality: What It Is, Why It Seems Scarce, Why It Matters* (Allen Lane, 2021)

Stuart Ritchie, *Science Fictions: Exposing Fraud, Bias, Negligence and Hype in Science* (Bodley Head, 2020).

Carl Sagan, *The Demon-Haunted World: Science as a Candle in the Dark* (Paw Prints, 2008)

Will Storr, *The Unpersuadables: Adventures with the Enemies of Science* (Overlook Press, 2014)

INDEX